Field Guide to the Cayuga Lake Region

Its Flora, Fauna, and Geology

Second Edition

James Dake

Foreword by Warren D. Allmon
Edited by Jonathan R. Hendricks

Paleontological Research Institution
Ithaca, New York
2018

Paleontological Research Institution
Special Publication No. 56

ISBN 978-0-87710-525-1
Library of Congress Control No. 2018949233

© 2018 Paleontological Research Institution
1259 Trumansburg Road
Ithaca, New York 14850 U. S. A.
http://www.priweb.org

This book was made possible through the generosity of the Triad Foundation.

First Printing of Second Edition: July 2018

On the cover: Spotted cucumber beetle (*Diabrotica undecimpunctata*) on a goldenrod (*Solidago* sp.). Photograph by Dayna Jorgenson.

TABLE OF CONTENTS

Foreword, by Warren D. Allmon 5
Introduction 7
Geology of the Cayuga Lake Region 11
Common Local Fossils 20
Flora
    Fungi and Lichens 24
    Ferns and Mosses 28
    Trees and Shrubs 30
    Wildflowers and Other Plants 44
Fauna
  Invertebrates
    Insects 64
        Butterflies and Moths 74
    Spiders and Kin 78
    Land Snails 82
    Other Freshwater and Land Invertebrates . . 86
  Vertebrates
    Amphibians 88
    Reptiles 94
    Birds 98
    Mammals 112
Photo Credits & Acknowledgments 121
Sources of More Information 123
Index 127

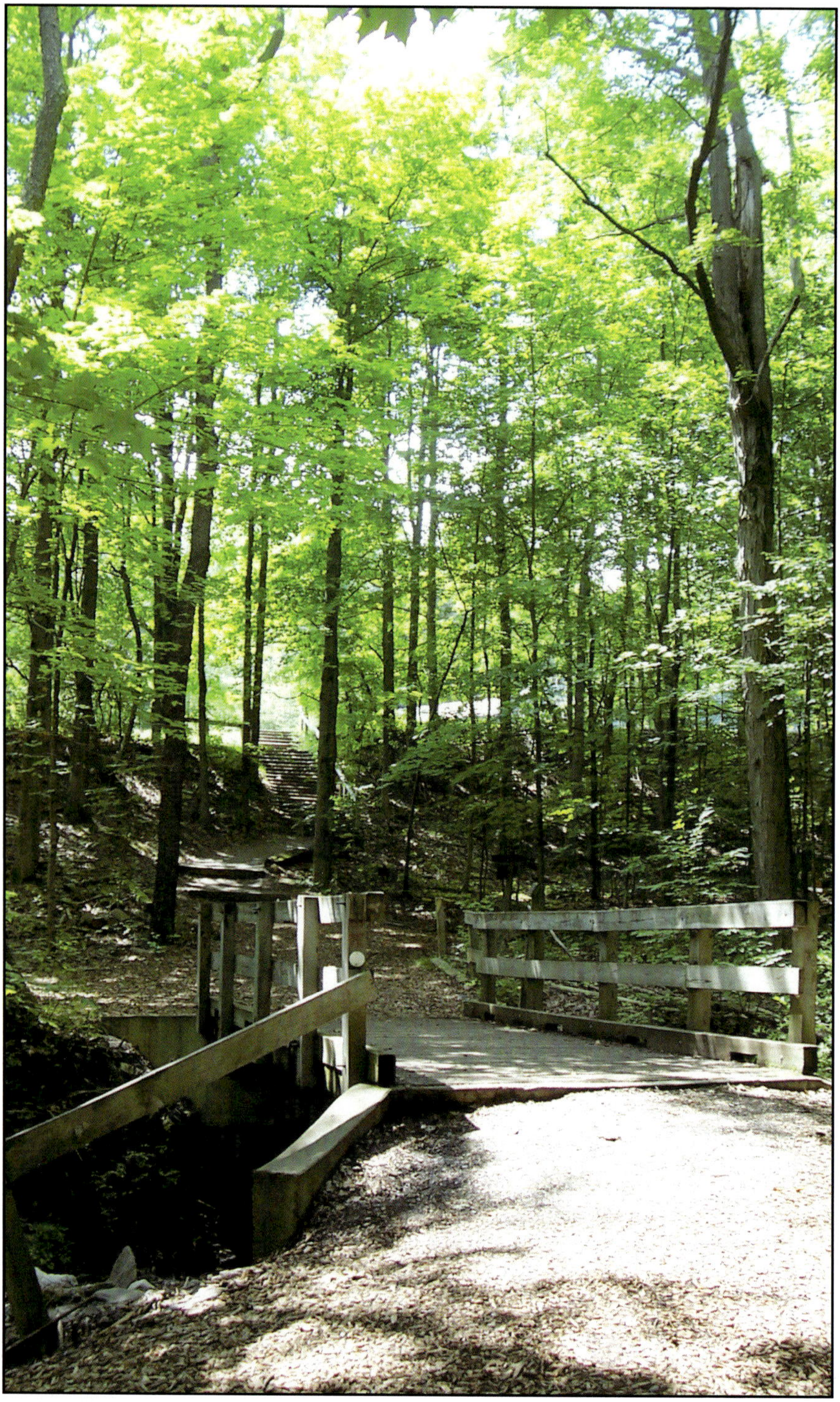

Foreword

This book is the descendant of two earlier landmark volumes by Ithaca authors. In 1911, Cornell professor Anna Botsford Comstock (1854-1930) published *Handbook of Nature Study*. Comstock was the founder and first head of the Department of Nature Study, and the first woman to be appointed to the Cornell faculty. Written originally for elementary school teachers, the book (which went through 24 editions and is still in print) was a gentle guide to everything that a student or parent or teacher might see in nature (at least the most common non-marine parts of it in the northeastern U.S.), how to observe it carefully, and why such study was important. Nature study, wrote Comstock, "consists of simple, truthful observations that may, like beads on a string, finally be threaded upon the understanding and thus held together as a logical and harmonious whole." Nature study, she contended, aims to cultivate the child's imagination and "love of the beautiful," aids "both in discernment and in expression of things as they are," and most importantly, "gives the child a sense of companionship with life out-of-doors and an abiding love of nature".

In 1949, Cornell professor of Nature and Science Education, Ephraim L. Palmer (1888-1970), published *Fieldbook of Natural History*. Palmer's book had much in common with Comstock's. It was written, he said, to address what he saw as an increasing gap between the technical literature of natural science and the "average person," who does not easily see how such science serves his daily life. "It is hoped," Palmer wrote in the Preface, "that this combination of philosophy, facts, and techniques may help us all enjoy doing what must be done, when it must be done wherever we may be. This should lead to a sound citizenship, a rational conservation policy, and a happy life."

Modern readers might find some of Comstock's and Palmer's lofty goals quaint. We no longer believe that nature "speaks" her truths if only we would hear them, nor do we share their Progressive-Era faith that life in the outdoors will cure all of society's ills. Yet the substance of "nature study" – as represented in these classic books and in the one you are now reading – is more important than ever. The reason for worrying about how modern life has increasingly isolated us from the natural world is not some hazy romantic notion about the purity of nature; it is because nature is rapidly changing under the relentless influence of human activity. We need to be aware of the plants and animals around us not just because they are pleasant and instructive, but because if we are not, they might not be here much longer. We need to understand nature to preserve what is left.

This book is thus both for the enjoyment and enlightenment of those fortunate enough to live in or visit the spectacularly beautiful but insufficiently appreciated Finger Lakes region. In using this guide, I hope that the reader will find that the beauty of this unique area lies not only in observing its life and landscape, but in the recognition that human understanding of nature is both its own reward and the key to our survival.

Warren D. Allmon, Director
Paleontological Research Institution

Cayuga Lake Watershed - Major Subwatersheds

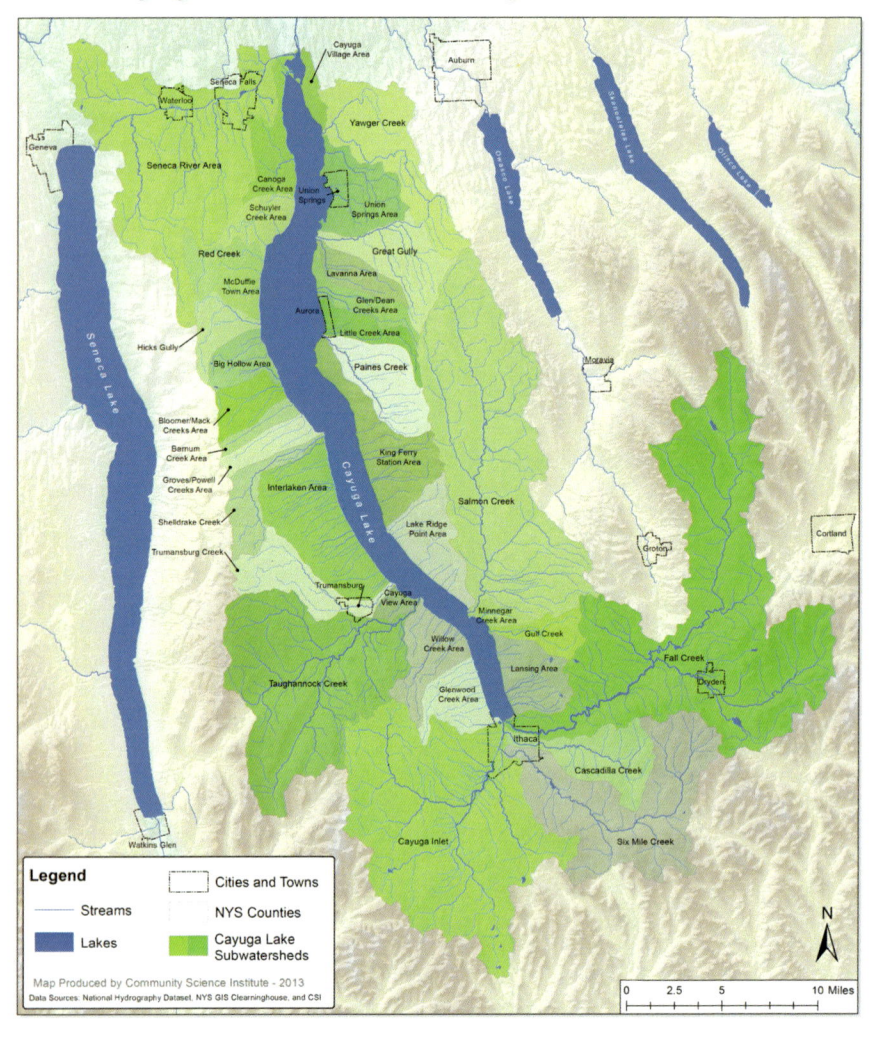

Map showing the boundaries of the Cayuga Lake Watershed. Image provided courtesy of The Community Science Institute (Ithaca, New York), which partners with volunteer groups to monitor water quality at over 100 stream locations throughout the Cayuga Lake watershed. Certified analytical results are shared free of charge at database.communityscience.org.

INTRODUCTION

The Cayuga Lake region, defined by the boundaries of the lake's watershed, abounds with a great diversity of plant and animal species. Many of these are commonly encountered during outdoor adventures in this beautiful area of Central New York. Native species that have made this region their home for thousands of years intermingle with recently introduced "alien" invaders from other parts of the world, all contributing to the modern ecology of the woods, fields, streams, and gorges surrounding Cayuga Lake.

This book is presented as a manual to identifying and appreciating the great diversity of present and ancient life found right here in Central New York. The goals of this book are to help you identify, learn about, and better appreciate more than 300 of these species—ranging from plants and fungi to insects, frogs, birds, and mammals—and the environments in which they live. Fossil remains of ancient sea life are also common in rocks exposed in gorges surrounding Cayuga Lake, recording the past life that inhabited the Finger Lakes region when a sea covered the area some 370 million years ago. This book will help you identify the types of fossils most commonly found in the local bedrock.

The complexity of the natural world is infinite. We could study this single piece of land for our entire lives, and still miss a lichen hidden in the crook of a branch, a fossil in an unturned rock, or a leaf miner burrowing its way through a single basswood leaf. Let this be a guide—a starting point—to study a single tier of the lands of the Cayuga Lake region. There will always be another layer of the natural world to peel back for those who wish to dig deeper. Look closely, and enjoy!

HOW TO USE THIS BOOK

This book is organized into three main sections: Geology, Flora, and Fauna. The flora and fauna sections are color-coded at the corners of the pages for easy location of the sections. The color photographs of each species or variety are located on the outside edges of the pages so that you can quickly find what you are looking for in the field. Side bars within each section inform you of characteristics of the species on the page you are examining. Each section has a brief introduction and an explanation of how that section is organized (*e.g.*, by color, size, or family). Each color photograph is accompanied by descriptive text for assistance in identification as below.

COMMON NAME
Scientific name, Family or Class
Descriptions of size, shape, colors, and traits for identification have **bold** headings to easily find what you are looking for.
Important characteristics to look for that distinguish a species from others are *italicized*.
Some species accounts note that they are alien or invasive. *Alien* species are exotic, or non-native. *Invasive* species are alien species that not only survive, but also flourish so well that they compete with or push out the species native to that area.

A close look at the stamens of Wild Columbine (Aquilegia canadensis)

Goldenrod amid a field of wildflowers

*A northern two-lined salamander (*Eurycea bislineata*) rests on a stream bed.*

Lucifer Falls at Enfield Glen, Robert H. Treman State Park

THE GEOLOGIC TIME SCALE

Era	Period	Millions of years ago
Cenozoic	Quaternary	Present — 2
Cenozoic	Tertiary	66
Mesozoic	Cretaceous	146
Mesozoic	Jurassic	200
Mesozoic	Triassic	251
Paleozoic	Permian	299
Paleozoic	Pennsylvanian	318
Paleozoic	Mississippian	359
Paleozoic	DEVONIAN	416
Paleozoic	Silurian	444
Paleozoic	Ordovician	488
Paleozoic	Cambrian	542
	Precambrian	4600

EARTH FORMED

10

GEOLOGY OF THE CAYUGA LAKE REGION

THE GEOLOGY OF NEW YORK STATE: A BRIEF SUMMARY

Approximately 600 million years ago, the Earth's continents were joined into one huge supercontinent, which geologists call Rodinia. This landmass began to break up and an ocean formed between two land masses that were roughly equivalent to what are today Europe and North America. This ocean continued to widen until around 500 million years ago, when spreading between the continents stopped, and the continents began to move back towards each other. The process of the continents coming together was complicated; smaller islands between the two continents rammed into what is today North America before Europe and Africa finally collided with it.

These collisions of landmasses pushed up a range of very high mountains along the eastern margin of North America. Today we call the remnants of those mountains the Appalachians, and they extend from eastern Canada to Alabama. Four hundred million years ago they were much higher than they are today.

For several hundred million years, global sea level was sufficiently high and the land surface west of the mountains was sufficiently low for the ocean to "flood" the region. This created a broad shallow sea covering much of the continent. At certain times (especially during the Silurian Period, 444-416 million years ago), this sea was very shallow indeed, and evaporation of seawater formed thick **salt deposits** that are mined today across central New York. One of the deepest salt mines in the world is located just northeast of Ithaca on the shore of Cayuga Lake in the town of Lansing. The mine produces rock salt for de-icing roads from a layer of salt approximately 2,500 feet (762 meters) below the surface.

Around 415 million years ago, sea levels rose. The area that is now the northeastern United States, including New York, was

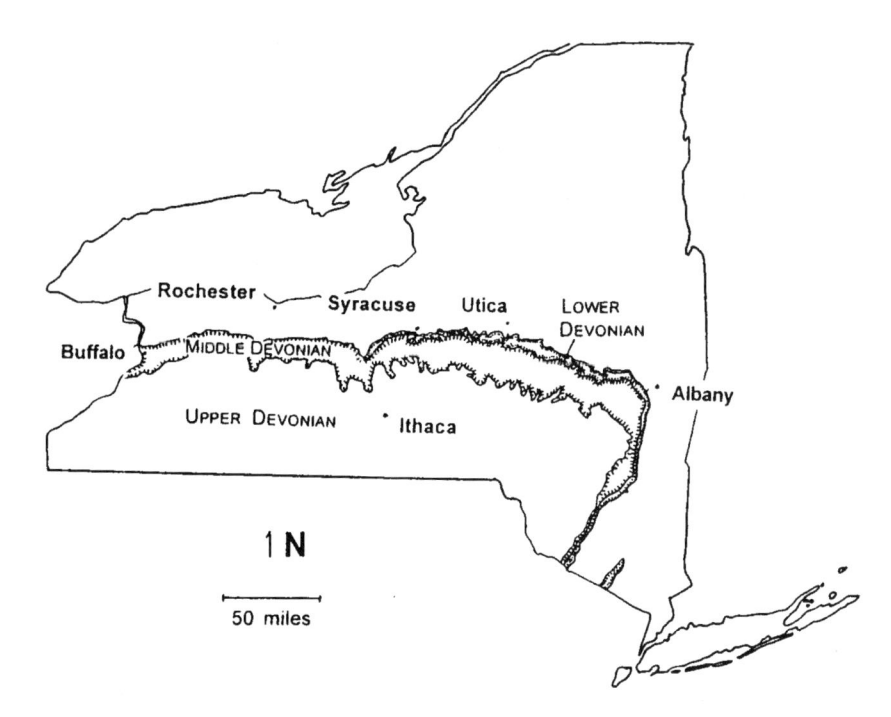

Devonian-aged rocks are at the surface over much of central and western New York State. The labeled areas on the map represent rocks from three intervals within the Devonian: Lower Devonian rocks are from the Early Devonian, about 400 million years ago (mya), Middle Devonian rocks are from about 380 mya, and Upper Devonian Rocks are from the Late Devonian (about 370 mya). Because the layers of rock in central New York are slightly tilted to the south, older layers are exposed to the north. Some Middle Devonian rocks outcrop to the south of the area shown in some deeper gorges (after Isachsen et al., 1991).

near the equator, and the oceans were warm and tropical. Marine life thrived on the sea bottom, and the skeletons of these marine organisms piled up on the seafloor. Some of these skeletons were made of calcium carbonate, which formed lime mud that eventually became **limestone**.

As the high mountains to the east of this sea continued to erode, huge quantities of gravel, sand, and mud flowed off the land into the shallow sea. These sediments formed what is now the thick stack of sedimentary rocks of Devonian age across much of central New York State. The coarse-grained sediments piled up in deltas and beaches close to land, forming **sandstones**. The fine-grained sediments stayed suspended in water longer and moved farther out to sea toward the west. They piled up in deeper, quieter waters and formed **siltstone** and **shale**. The blackest mud accumulated in water that contained little or no oxygen (and so the black carbon in the mud did not oxidize); this mud today forms the great thicknesses of dark gray and black shales throughout much of central and western New York.

By the end of the Devonian Period (359 million years ago), as deltas from the eroding mountains built out into central New York and as global sea levels fell, the shallow seas began to retreat from what is now the northeastern United States. The sea continued to exist for a time in other parts of the continent; the great coal beds of Pennsylvania resulted from coastal forests that covered this area during the Carboniferous Period, immediately after the Devonian.

By the start of the **Age of Dinosaurs** around 240 million years ago, all of the continents were once again joined into one large supercontinent, which geologists call Pangaea, and the region that is now the northeastern U.S. was dry land. During that time, dinosaurs left their footprints along rivers and lakes at the bottom of "rift" valleys, where Pangaea began to split apart again. These fossils are abundant in the red sandstones of the Connecticut River Valley of Massachusetts and Connecticut, but such sandstones are only sparsely represented in the southeastern corner of New York State. Only eight sets of small fossil footprints are known from Nyack, in Rockland County. Any other rocks of this age that might have been present elsewhere in New York have eroded away. You can see one of these New York dinosaur footprints on exhibit at the Museum of the Earth in Ithaca.

Even though we have almost no direct evidence, there were almost surely dinosaurs wandering through central New York for many millions of years. After the dinosaurs became extinct (around 66 million years ago), mammals (animals with fur and mammary glands) occupied the landscape. But these events are not recorded in central New York, because, as for dinosaur-bearing rocks, sediments and fossils within them have been long erased, weathered, and eroded by winds and rain, freezing, and thawing, and finally bulldozing by sheets of ice. This erosion has even erased many of the uppermost rocks from the Devonian Period.

The next recorded chapter of the geological history of New York State is the Ice Age, which began approximately two million years ago, and left the lakes, gorges, and gravel of the Finger Lakes region and elsewhere behind as a reminder. In the landforms and glacial gravel we find a record of how the glaciers moved, and sometimes find the fossil bones of animals, such as **mastodons**, which were distant relatives of elephants that lived along the margin of the great sheets of ice. At least 130 mastodon skeletons have been found in New York State over the past 300 years, including in Tomkins County where Ithaca is located. One of the most complete mastodon skeletons ever found is the Hyde Park Mastodon, discovered in 1999 in Dutchess County in the Hudson River Valley. You can see this 13,000-year-old skeleton mounted at the Museum of the Earth in Ithaca.

The Hyde Park Mastodon in the Quaternary world exhibit at the Museum of the Earth. Photograph by Rachel Philipson, 2008.

Formation of the Finger Lakes

The Finger Lakes consist of 11 long, narrow, roughly parallel lakes, oriented north-south like fingers of a pair of outstretched hands. The southern ends have high walls, cut by steep gorges. Two of the lakes (Seneca and Cayuga) are among the deepest in North America and have bottoms below sea level. These lakes all formed over the last two million years by glacial carving of old stream valleys. Ithaca is located at the southernmost end of Cayuga Lake, longest and second deepest of the Finger Lakes. Cayuga is 38.1 miles (61.3 kilometers) long and 435 feet (133 meters) deep (53 feet or 16 meters below sea level) at its deepest spot. The actual depth of carved rock is well over twice as deep, but it has been filled with sediments; there could be as much as 1000 feet (300 meters) of glacial sediment in the deep rock trough below the lake bed. Seneca Lake to the west is more than 600 feet (180 meters) deep at its deepest point.

The Finger Lakes originated as a series of northward-flowing rivers that existed in what is now central New York State (today most of central New York drains to the south). Around two million years ago the first continental glaciers moved southward from the Hudson Bay area, initiating the Pleistocene glaciation, commonly known as the Ice Age. The Ice Age was really a series of many advances of glaciers (one approximately every 100,000 years). The advances are called **glacials**, and the retreats are called **interglacials**. Today we live during an interglacial. The Finger Lakes were probably carved by several of these episodes. Ice sheets more than

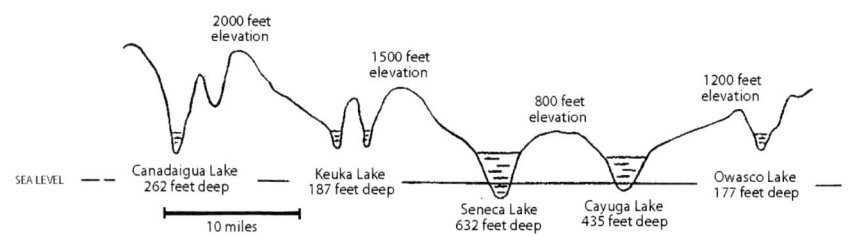

A west-to-east cross section of the larger Finger Lakes (after Figiel, 1995).

2 miles (3 kilometers) thick flowed southward, parallel but opposite to the flow of the rivers, gouging deep trenches into these river valleys. Traces of most of the earlier glacial events have been eroded away, but much evidence remains of the last one or two glacials that covered New York.

The latest glacial episode was most extensive around 21,000 years ago, when glaciers covered almost the entire state. Around 19,000 years ago, the climate began to warm, and the glaciers began to retreat, disappearing entirely from New York most recently around 11,000 years ago.

One obvious bit of evidence left by the glaciers is the gravel deposits at the southern ends of the Finger Lakes, called **moraines**, which are piles of rock pushed along the edge of the advancing glacier. Another are teardrop-shaped hills of glacial sediment called **drumlins**, which formed under the

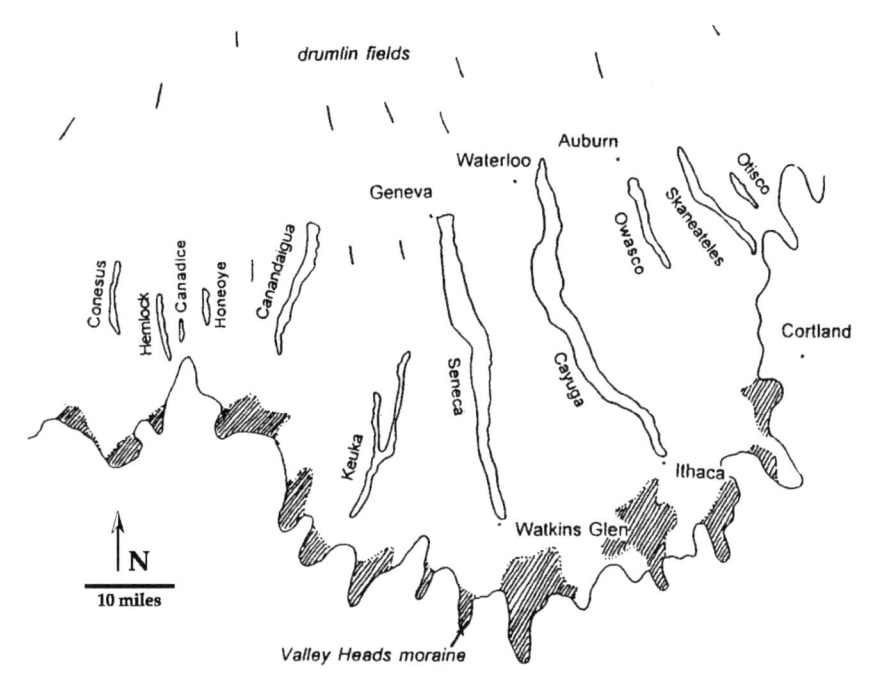

The Finger Lakes. Glacial gravel from the edge of a retreating glacier can be found in the Valley Heads moraines (hatched areas). Short lines to the north represent orientation of teardrop-shaped hills of glacial gravel called drumlins, found especially between Syracuse and Rochester.

glacier as it advanced. Moraines are visible south of Ithaca at North Spencer, along Route 13 west of Newfield, and northeast of Ithaca between Dryden and Cortland. Drumlins are visible at the northern ends of Cayuga and Seneca lakes in a broad band from Rochester to Syracuse.

The Devonian and glacial history of the Finger Lakes region are also responsible for the wine industry of the region. More than 50 vineyards are scattered around the central Finger Lakes, usually on the sloping sides of the lake valleys. The fractured shale and limestone enable the long roots of grape vines to penetrate deeply into the ground, and also keep the overlying soils well-drained. The topography of the lake valleys provide bands of appropriate climatic conditions (called microclimates), with temperatures and humidities appropriate for growing several European vinifera varieties. The lakes temper the upstate winters, often keeping vineyards considerably warmer than locations less than a mile away, and also moderate the transitions of spring and fall.

"ITHACA IS GORGES" — WHY?

The famous gorges of the Finger Lakes are a result of the interaction between the south-to-north-running river valleys, which were gouged by glaciers numerous times over the last 2 million years, and streams running perpendicular to the glaciers. The story of the gorges began when the Finger Lakes

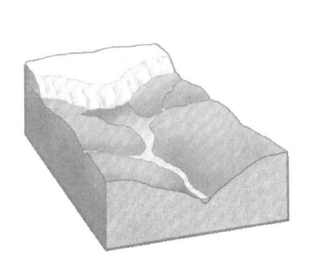

As glacial ice repeatedly advanced and retreated over the Finger Lakes river valleys (left) it gouged deep trenches into these valleys (right). Illustrations by J. Houghton.

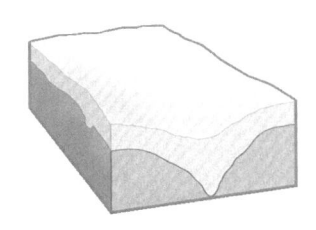

were river valleys with small streams flowing in from the east and west. The valleys were repeatedly filled with hundreds of feet of glacial ice that originated from glaciers flowing south out of Canada, eroding the valleys more deeply.

When the glaciers melted back, and glacial sediment dammed the river valleys, lakes much deeper than those of today formed and streams plunged over waterfalls from the glacially steepened hills. As the lake levels dropped, a series of steps were left on the hillsides, like those seen at the top of Taughannock Falls. The streams eroded downward, forming the gorges so characteristic of the Finger Lakes region, in a process that continues today. The material that the streams eroded was deposited in large deltas at the mouths of the creeks. These deltas are clearly visible today at places such as Taughannock Falls State Park and Myers Point, near the southern end of Cayuga Lake. Some of the gorges were cut during earlier interglacials, filled with glacial sediment during ice advances, and then re-cut since the last glacial retreat.

Erosion of the gorges appears today to be mostly slow and gradual. There are rounded pebbles worn smooth by the water and occasionally rounded holes in the stream beds (called plunge pools and potholes) that have been scoured out by stones whirled around by the water. But things are not always as they seem. The flow of streams in upstate New York is highly seasonal, with high volume in the spring from snowmelt and low volume in the summer and fall. More erosion is likely to happen when there is more water flowing in the stream.

Look carefully at the rocks themselves and you will see other signs that erosion is not always constant and gradual. The rocks around Ithaca are cut by thousands of ruler-straight cracks, which look like they have been cut with a saw. These are natural fractures called **joints**. They are caused by stress of rocks on an enormous geographic scale, due to the collision of the continents more than 250 million years ago. These joints form weaknesses in the rocks of the gorge walls. Water seeps into the cracks, freezes, and expands. Eventually, catastrophic failure occurs, resulting in a rock

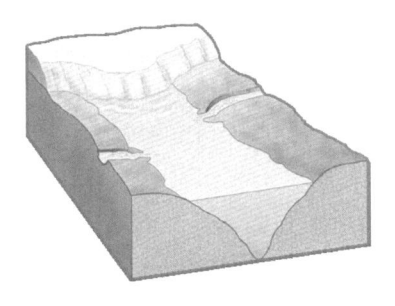

As glacial ice melted, sediment dammed the Finger Lakes river valleys, forming deep lakes. Streams plunged into the lake from the glacially steepened valley sides.

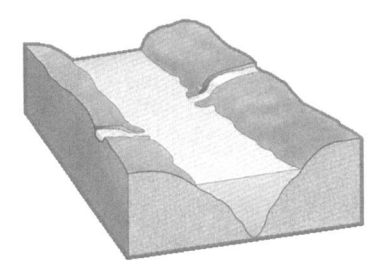

Later, the lake level fell, leaving a series of steps on the hillsides.

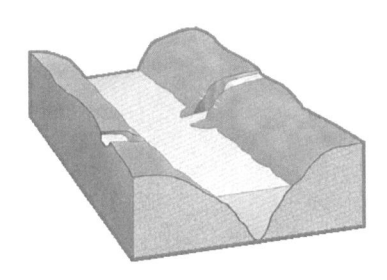

Streams then eroded downward as the lake level continued to drop, forming gorges and deltas.

Illustrations by J. Houghton.

slide. The broken rocks are then moved downstream by spring floods and eventually out into the main valley or lake. Look at the fracture patterns in the walls of the gorges. Look at the piles of rocks at the bases of the walls. You can see that the gorges formed by this system of small catastrophes and variation of flow in the streams.

The previous three sections were reprinted with permission from *Ithaca is Gorges, 4th ed.*, by Allmon & Ross, Paleontological Research Institution, 2008.

COMMON LOCAL FOSSILS

Many of the fossils found in this area are located in the fossil-rich band of Middle Devonian rocks stretching across New York State north of Ithaca. In this section, information is included about each of the main groups of organisms commonly fossilized in the local bedrock. By knowing the age range in which fossils of these animals have been found, geologists are able to relatively date the rocks in this area. Also, by knowing the environment that these organisms lived in (by comparison with similar animals living today), they are able to understand the environmental conditions at the time that these rocks were formed.

TRILOBITES
Phylum Arthropoda (arthropods)
Subphylum Trilobita
Age Range Cambrian to Permian, most common from Cambrian to Devonian. **Identification** arthropods, related to horseshoe crabs and lobsters, segmented bodies, some with compound eyes, fragments of molted skins are most often found. **Environment** marine.

BRACHIOPODS
Phylum Brachiopoda
Age Range Late Proterozoic to Recent.
Identification most common fossil in the region, two shells made of calcium carbonate hinged together, each individual shell *symmetrical from side to side*. **Environment** marine.

BRYOZOANS
Phylum Bryozoa
Age Range Late Proterozoic to Recent, most common from Ordovician to Permian. **Identification** colonial, made of calcium carbonate, somewhat resembling corals, can be branched or web-like. **Environment** shallow tropical marine.

CORALS
Phylum Cnidaria (corals)
Order Rugosa (horn corals), Order Tabulata (tabulate corals)
Age Range Ordovician to Permian. **Identification** colonial, composed of calcium carbonate, horn- or cornucopia-shaped, some with honeycomb or chain-like patterns. **Environment** shallow tropical marine.

CRINOIDS
Phylum Echinodermata (echinoderms)
Class Crinoidea (sea lilies)
Age Range Ordovician to Recent, most common from Ordovician to Permian. **Identification** multi-armed head (calyx) on stem, related to starfish and sea urchins, usually fossilized as broken stem fragments that look like *small circular discs*. **Environment** shallow marine.

BIVALVES
Phylum Mollusca (mollusks)
Class Bivalvia (clams)
Age Range Cambrian to Recent, most common from Triassic to Recent. **Identification** clams, two shells made of calcium carbonate hinged together, each individual shell symmetrical to its counterpart. **Environment** aquatic, mostly marine, some freshwater.

FOSSILS

CEPHALOPODS
Phylum Mollusca (mollusks)
Class Cephalopoda (squids, octopuses)
Age Range Cambrian to Recent, most common from Ordovician to Cretaceous. **Identification** straight-shelled nautiloids with sectioned chambers that look like a drill bit, or ammonoids with a curved spiraling shell with complex chambers. **Environment** deep marine.

GASTROPODS
Phylum Mollusca (mollusks)
Class Gastropoda (snails)
Age Range Cambrian to Recent, most common from Cretaceous to Recent. **Identification** snails, small coiled or spiraled shells. **Environment** marine, freshwater, or land.

FOSSILS

The Tully Limestone is exposed at the lower falls at Taughannock Falls State Park. This resistant sedimentary rock layer forms the stream bed and is much stronger than the over- and underlying shale layers.

Flat joint surfaces (arrows) are common rock features in Central New York. One of the best places to see them is at Enfield Glen in Robert H. Treman State Park. In some locations, joints control the paths of creeks and streams, resulting in sharp, nearly 90 degree turns.

FUNGI AND LICHENS

The **Fungi** Kingdom includes mushrooms, molds, yeasts, and mildews. Fungi lack chlorophyll, and so must obtain food in other ways. Some fungi are **saprophytic** and break down dead plants or animals. Others are **parasitic** and feed off of living plants or animals. Others have a symbiotic (mutually beneficial) relationship with the roots of trees or shrubs and are called **mycorrhizal**. Fungi reproduce using spores. Many fungi often change drastically in appearance over time, and so are very difficult to identify.

Lichens are dual organisms made up of a fungus surrounding green algae or cyanobacteria. The fungus takes water and minerals from its surroundings, whereas the algae uses photosynthesis to make food. Lichens reproduce by spores, dispersing small packets of algae and fungus, or through fragments of the lichen that break off to form new ones. Lichens can live in many different environments and grow very slowly.

FUNGI

SULFUR SHELF
Laetiporus sulphureus, Polypore Family
Width 7-10". **Color** bright orange-yellow. **Description** shelf-like or rose-shaped, overlapping, tough with soft edges, stalk-less, gill-less. **Season** spring, summer, and fall. **Found** on living or dead trees, parasitic or saprophytic.

TURKEY TAIL
Trametes versicolor, Polypore Family
Width 1-2". **Color** variable, brown with whites, yellows, blues, and greens. **Description** *velvety*, tough, *flexible*, will bend without breaking, stalk-less. **Season** late spring to fall. **Found** usually on dead hardwoods, saprophytic.

STUMP PUFFBALL
Lycoperdon pyriforme
True Puffball Family
Width ½-2". **Color** off white to brown, gleba (spores inside puffball) white turning brown. **Description** pear-shaped, finely spiny, bursts releasing spores when disturbed by touch or rain or wind, unpleasant odor. **Season** late summer, but can be found year-round. **Found** on dead hardwood, usually in clusters, saprophytic.

CINNABAR CHANTERELLE
Cantharellus cinnabarinus
Chanterelle Family
Width 1½". **Color** bright red-orange with some white. **Description** stalk 1½" with irregular-edged lobed cap, dry, hairless, with folds on underside. **Season** summer to fall. **Found** on ground, mycorrhizal with hardwoods.

WITCH'S HAT
Hygrocybe conica, Wax Cap Family
Width ½-2½". **Color** red to orange, bruises black. **Description** stalk 2-4", hollow, twisted, with white base, cap rounded with narrow point, hairless. **Season** spring to fall. **Found** on ground under hardwoods or conifers, saprophytic.

WITCH'S BUTTER
Tremella mesenterica, Jelly Fungi Class
Width 1-4". **Color** orange to yellow. **Description** lobed, folded or brain-like, gelatinous but tough when older, stalk-less. **Season** late summer to fall, usually after rain. **Found** on dead hardwoods, parasitic on other fungi.

FUNGI

GRAY URN FUNGUS
Urnula craterium
Large Cup Fungi Family
Width 1-2". **Color** dark brown to black. **Description** urn-shaped, 3-4" tall, tough, closed at first and then split open. **Season** spring to summer. **Found** on dead hardwoods, usually buried, saprophytic.

COMMON GREENSHIELD
Flavoparmelia caperata, Lichen Group
Width 1-4". **Color** pale green, underside black. **Description** rounded spreading lobes in rosettes, wrinkled-looking, covering bark. **Season** year-round. **Found** usually on branches and trunks of trees, photosynthetic.

BRITISH SOLDIERS
Cladonia cristatella, Lichen Group
Height ¾". **Color** gray-green with red tops. **Description** stalks scaly, erect, often numerous in groups, each with red, rounded fruiting body on top. **Season** year-round. **Found** on wood or soil, photosynthetic.

GIANT ROCKTRIPE
Umbilicaria mammulata, Lichen Group
Width 2-5". **Color** brown, underside black. **Description** discs with folded rippled sides, leathery and smooth, hard when dry, hairs on underside. **Season** year-round. **Found** on rock and boulders in open forests, photosynthetic.

A variety of colorful fungi growing along the trunk of a fallen Black Birch. Fungi can be found year round, but most often after heavy rains on decaying plant matter.

FERNS AND MOSSES

Ferns are seedless vascular plants with a rhizome that grows beneath the ground and bears roots and leaves (fronds). Each spring, the young fronds (called fiddleheads when they first sprout) unfurl from the rhizome; they are usually compound and divided giving them a feathery appearance. Some ferns are evergreen, keeping their fronds year round. Ferns reproduce by spores located in tiny sacs called sporangia. These are usually clustered on the underside of the fronds in different shapes called sori. Ferns are usually found in moist, shady areas in forests.

Mosses are small green plants often found growing on rocks, trees, logs, or the ground in damp low-light environments. They reproduce by spores released from a tiny capsule on a stalk that is visible when they are "fruiting," and also by breaking apart or sending out new shoots. They lack a well-differentiated internal vascular system and so absorb and move water and nutrients largely through external means.

NEW YORK FERN
Thelypteris noveboracensis
Marsh Fern Family
Length 12-24". **Fronds** green-yellow, twice compound, 20 or more pairs of leaflets, tapering at tip *and at base*. **Sori** rounded on underside leaflet edges. **Found** in clearings in moist woods.

SENSITIVE FERN
Onoclea sensibilis, Wood Fern Family
Length 24-36". **Fronds** pale green, once compound, approximately 12 pairs of leaflets with wavy edges, leathery, dying after first frost giving this fern its name. **Sori** on separate fertile frond, clustered, bead-like, brown, persists through winter. **Found** in moist woods.

CHRISTMAS FERN
Polystichum acrostichoides
Wood Fern Family
Length 18-30". **Fronds** evergreen, once compound, 20 to 40 leaflets, usually alternating, finely toothed and uneven at base. **Sori** rounded, covering undersides of upper leaflets. **Found** near streams.

SPINULOSE WOOD FERN
Dryopteris carthusiana
Wood Fern Family
Length 12-28". **Fronds** green, thrice divided, leaflets toothed with up to 18 pairs, lowest inner leaflets longer than others, on tan scaly stalks. **Sori** rounded in rows on underside of leaflets. **Found** in moist woods.

WHITE CUSHION MOSS
Leucobryum glaucum
True Mosses Class
Length 1-3", grows upright (acrocarp). **Leaves** several cell-layers thick with layers of dead cells (hyaline cells), can store 400x of moss's dry weight in water, hair-like, pale green to blue-green, growing in clump or ball up to 2' or more, sometimes mat-like. **Capsule** curved, red-brown, on thin stalk. **Found** on forest floor, over rotting wood and moist areas.

ROUGH-STALKED FEATHER MOSS
Brachythecium rutabulum
True Mosses Class
Length 1-6", grows horizontally (pleurocarp). **Leaves** glossy, pale green, ovate, pointed, with midvein (costa) present. **Capsule** curved, smooth, dark red-brown on rough stalk. **Found** on logs, leaf litter, and bases of trees.

MOSSES

29

TREES AND SHRUBS

This section covers **woody plants** that do not die back each year, but continue to grow larger annually as seen in their annual growth rings. Species are organized first by the tree type: **Conifers** (*gymnosperms*, having exposed seeds, with needle or scale-like leaves covered in a waxy coating), **Broadleaf Trees** (*angiosperms*, flowering and fruiting trees, with broad flat leaves that are mostly *deciduous*, meaning they shed their leaves for winter), and **Small Trees and Shrubs** (broadleaf trees that generally grow to 20 feet or less).

The broadleaf trees are organized by the **arrangement of their leaves** (alternate or opposite, simple or compound), and finally by **family**.

CONIFERS

EASTERN RED CEDAR
Juniperus virginiana, Cypress Family
Bark dry, shreddy, red-brown. **Leaves** scale-like, covering *rounded* twigs.
White Cedar (*Thuja occidentalis*) has *flat* scale-like leaves. **Cones** small, blue, and berry-like. **Try** scratching and smelling the bark, which has rich cedar scent.

NORWAY SPRUCE
Picea abies, Pine Family
Bark scaly red-brown, *drooping branches*. **Needles** ¾", dark green, single, four-sided like all spruces.
Cones 5" long, green turning brown, stiff scales. Introduced from Europe.
Try searching below the tree to find its long slender cones that fall soon after maturity.

30

WHITE SPRUCE
Picea glauca, Pine Family
Bark scaly gray-brown. **Needles** 1/3-¾", blue-green, 4-sided, single, stiff. **Cones** 1-2", flexible scales. **Try** crushing a few of the needles in your hands and smell why this tree also gets the name, "Skunk Spruce."

RED SPRUCE
Picea rubens, Pine Family
Bark scaly red-brown to gray. **Needles** ½", green-yellow, 4-sided, single, sharp, curved upward. **Cones** 1½", red-brown. **Try** crushing a few of the needles in your hands and smell the almost citrus scent.

RED PINE
Pinus resinosa, Pine Family
Bark red-brown, scaly. **Needles** 4-6", in groups of two, dark green. **Cones** 2", egg-shaped, green turning brown. **Try** bending the paired needles of this tree. They should break cleanly in half.

SCOTCH (SCOTS) PINE
Pinus sylvestris, Pine Family
Bark orange-brown, scaly, *bright orange on upper trunk* and branches. **Needles** 2-3", in groups of two, twisted, green-blue. **Cones** 1½", brown. Introduced from Europe. **Try** looking for the bright orange bark and shorter needles that distinguish this tree from the Red Pine.

CONIFERS

31

BROADLEAF TREES: alternate simple

EASTERN WHITE PINE
Pinus strobus, Pine Family
Bark gray-black, scaly. **Needles** 4", soft, flexible, blue-green, in groups of five. **Cones** 6", curved, resinous. **Try** feeling the uniquely long soft needles that grow in bundles of five.

EASTERN HEMLOCK
Tsuga canadensis, Pine Family
Bark gray-brown turning red-brown, scaly. **Needles** ½", green, white beneath, *flexible, flat*. **Cones** ¾", tiny, few scales, hang at twig ends. **Try** finding the very small cones on this tree.

AMERICAN BEECH
Fagus grandifolia, Beech Family
Bark *smooth* and gray. **Leaves** alternate simple, 4", coarse-toothed, green, football-shaped, with distinct straight veins. In autumn, yellow-brown, often stay on tree. **Fruit** ¾" prickly brown bur, nuts eaten by many animals. **Try** feeling the bark which some compare to the skin of an elephant.

WHITE OAK
Quercus alba, Beech Family
Bark furrowed, gray. **Leaves** alternate simple, 3-9", 7-11 *rounded lobes*, hairless, green, whitened below. In autumn, dull red-bronze. **Fruit** acorns, with bumpy rounded bowl, 1¼".

BLACK OAK
Quercus velutina, Beech Family
Bark dark gray-black, blocky, furrowed. **Leaves** alternate simple, 4-10", 7-11 *deep pointed* lobes, leathery, *shiny green*. In autumn, dull red-bronze. **Fruit** acorns, with shaggy bowl, ¾".

NORTHERN RED OAK
Quercus rubra, Beech Family
Bark dark gray, furrowed. **Leaves** alternate simple, 4-10", 7-11 *shallow pointed* lobes, thin, *dull* green. In autumn, dark red. **Fruit** acorns, with flat scaly bowl, 1". **Try** finding the different acorns of each oak species.

PAPER (WHITE) BIRCH
Betula papyrifera, Birch Family
Bark white, *peeling*, brown underneath. **Leaves** alternate simple, 1-4", triangular, toothed. In autumn, light yellow. **Fruit** 1¾" brown drooping slender flower cluster. **Try** feeling the dry papery bark of this tree.

BLACK (SWEET) BIRCH
Betula lenta, Birch Family
Bark black, slightly peeling, horizontal stripes. **Leaves** alternate simple, 1-6", tear-shaped, toothed, with forked veins. In autumn, bright yellow. **Fruit** 1", upright brown flower clusters. **Try** crushing a leaf or breaking a twig to smell the wintergreen oil in this tree's sap.

BROADLEAF TREES: alternate simple

BROADLEAF TREES: alternate simple

AMERICAN HORNBEAM (MUSCLEWOOD)
Carpinus caroliniana, Birch Family
Bark gray, *smooth, rippled*, muscular-looking. **Leaves** alternate simple, 1-5", tear-shaped, double-toothed, veins parallel, *non-forked*. In autumn, orange-scarlet. **Fruit** 3" three-pointed leafy clusters with tiny nuts.

EASTERN HOP HORNBEAM (IRONWOOD)
Ostrya virginiana, Birch Family
Bark brown-gray, *shreddy, vertical lines*. **Leaves** alternate simple, 1-5", tear-shaped, double-toothed, veins parallel and *forked*. In autumn, yellow. **Fruit** 2" hanging cluster of sacs with small nuts.

SLIPPERY ELM
Ulmus rubra, Elm Family
Bark dark gray-brown furrowed, showing *orange inner bark*. **Leaves** alternate simple, football-shaped with uneven base, 4-8", *thick, rough*, double-toothed. In autumn, yellow. **Fruit** ½", circular, slightly notched pouches. **Try** feeling the rough, almost sandpapery, leaves of this tree.

AMERICAN BASSWOOD
Tilia americana, Linden Family
Bark gray, tough, smooth, shallow grooves. **Leaves** alternate simple, 3-10", heart-shaped with *uneven base*, toothed. In autumn, yellow. **Fruit** ½" nuts clustered under leafy wing. **Try** smelling the fragrant yellow flowers that bloom from June to August.

CUCUMBER MAGNOLIA
Magnolia acuminata, Magnolia Family
Bark dark brown, furrowed. **Leaves** alternate simple, 4-10", egg-shaped, shiny. In autumn, yellow-red. **Fruit** cucumber-shaped cone with red seeds. **Try** looking for the yellow-green cup-shaped flowers that bloom from May to June.

TULIP TREE
Liriodendron tulipifera, Magnolia Family
Bark light gray with white grooves. **Leaves** alternate simple, 6-10", four-pointed lobes. **Fruit** large tulip-like orange and green flowers giving way to 1-2" winged seeds clustered in 3" cone. **Try** finding the large tulip-like orange and green flowers from May to June.

APPLE
Malus pumila, Rose Family
Bark brown-gray scaly. **Leaves** alternate simple, 1-4", egg-shaped, round-toothed, hairy beneath. In autumn, yellow. **Fruit** greater than 1" across, yellow to red. Introduced from Eurasia.

PEAR
Pyrus communis, Rose Family
Bark gray-brown scaly. **Leaves** alternate simple, 1-3", egg-shaped, round-toothed, nearly hairless. In autumn, yellow. **Fruit** 3-4" green. Introduced from Eurasia. **Try** looking for rows of holes made in the bark of this tree by Yellow-bellied Sapsuckers.

BROADLEAF TREES: alternate simple

BROADLEAF TREES: alternate simple

BLACK CHERRY
Prunus serotina, Rose Family
Bark dark gray, flaky, scaly, short horizontal lines, red-brown under bark. **Leaves** alternate simple, 2-6", football-shaped, toothed, glossy, creased along midrib. In autumn, yellow-red. **Fruit** ½" black-red berries hanging in 5" clusters. **Try** feeling the flaky bark, which some naturalists say looks like burnt potato chips.

WEEPING WILLOW
Salix babylonica, Willow Family
Bark gray, furrowed. **Leaves** alternate simple, 1-5", narrow and pointed, fine-toothed, white beneath, drooping on *long sweeping branches*. In autumn, yellow. Introduced from Eurasia.

EASTERN COTTONWOOD
Populus deltoides, Willow Family
Bark gray, *rough*, rigid. **Leaves** alternate simple, 2-8", triangular with long flatted leafstalks, toothed, shiny. In autumn, yellow. **Fruit** ½" brown capsules in 8" hanging clusters. **Try** looking for the cottony seeds floating through the air in late spring as they are released from their capsules.

BIGTOOTH ASPEN
Populus grandidentata, Willow Family
Bark smooth brown-yellow, shallow furrows. **Leaves** alternate simple, 2-6", triangular, rounded, with long flattened leafstalks, *coarse-toothed*. In autumn, yellow-red. **Fruit** 2½" cluster of green capsules with cottony seeds.

QUAKING ASPEN
Populus tremuloides, Willow Family
Bark smooth white-yellow. **Leaves** alternate simple, 2-4", rounded with point, long flattened leafstalks, *fine-toothed*. In autumn, bright yellow. **Fruit** 4" hanging clusters of narrow green capsules with cottony seeds. **Try** feeling the flat leafstalks that make this tree's leaves "quake" in the slightest wind.

NORTHERN CATALPA
Catalpa speciosa, Bignonia Family
Bark gray-brown, scaly. **Leaves** opposite simple, 6-13", heart-shaped. In autumn, yellow-brown. **Fruit** seed pods, 10-24". **Try** looking for the white trumpet-shaped purple and yellow fringed flowers in May-June.

RED MAPLE
Acer rubrum, Maple Family
Bark dark gray, broken. **Leaves** opposite simple, 2-8", 3-5 lobes, *fine-toothed* along edges, white beneath. In autumn, red-orange. **Fruit** 1" paired seeds with wings, reddish.

SUGAR MAPLE
Acer saccharum, Maple Family
Bark dark brown-gray, furrowed, blocky. **Leaves** opposite simple, 2-10", mostly 5 lobes, pointed, firm. In autumn, orange-yellow, red. **Fruit** 1¼" paired seeds with wings. **Try** breaking a leaf stem to see the clear sap. The similar European **Norway Maple** (*A. platanoides*), often planted in neighborhoods, has milky white sap.

BROADLEAF TREES: opposite simple

BROADLEAF TREES: alternate compound

BLACK LOCUST
Robinia pseudoacacia, Pea Family
Bark gray, deeply ridged. **Leaves** alternate compound, 6-12", 6-20 1½" rounded oval leaflets. **Honey Locust** (*Gleditsia triacanthos*) has narrower leaflets. In autumn, yellow. **Fruit** 6" brown-black flat hanging seed pods. **Try** smelling the hanging white flowers during May to June that have a grape-like scent.

SHAGBARK HICKORY
Carya ovata, Walnut Family
Bark light gray, *peeling in shaggy loose vertical strips*. **Leaves** alternate compound, 8-14", 5-7 5" football-shaped leaflets, toothed. In autumn, golden. **Fruit** 2½" green-brown four-angled husked nuts.

BLACK WALNUT
Juglans nigra, Walnut Family
Bark dark gray, deeply grooved. **Leaves** alternate compound, 12-24", 7-17 narrow football-shaped leaflets, toothed, hairy beneath, end leaflet small or lacking. In autumn, yellow. **Fruit** 2½" husked shelled nuts, round, green-brown. **Try** smelling the spicy scent of the crushed leaves.

BOX ELDER
Acer negundo, Maple Family
Bark brown-gray, furrowed. **Leaves** opposite compound, 4-10", with 3-7 4" leaflets, coarse-toothed or smooth, some with lobes. In autumn, yellow. **Fruit** 1½" paired seeds with wings, but drooping. **Try** finding a Box Elder sapling, which sometimes resembles poison ivy. Poison ivy will grow alternate along the stem rather than opposite.

WHITE ASH
Fraxinus americana, Olive Family
Bark dark gray, tightly furrowed into forking crossed ridges. **Leaves** *opposite* compound, 8-12", with 5-9 4" football-shaped leaflets, toothed or not, *whitish underneath*. In autumn, purple-yellow. **Fruit** 1½" narrow hanging winged seeds.

GREEN ASH
Fraxinus pennsylvanica, Olive Family
Bark gray, tightly furrowed into forking crossed ridges. **Leaves** *opposite* compound, 8-10", with 5-9 4" football-shaped leaflets, toothed or not, green on both sides, *tiny hairs on underside*. In autumn, yellow. **Fruit** 2" narrow hanging winged seeds.

PAWPAW
Asimina triloba, Annona Family
Bark dark brown, smooth. **Leaves** alternate simple, 6-12", light green, tear-shaped, wider towards end, toothless. **Fruit** 4", green, banana-like. **Try** looking for the maroon flowers that bloom in April or May, before the leaves.

HIGH-BUSH BLUEBERRY
Vaccinium corymbosum, Heath Family
Bark yellow-green to red on many shrubby stems. **Leaves** alternate simple, 1-3", rounded, waxy above, some hair beneath. In autumn, bronze-red. **Flowers** ¼", white, bell-shaped, May—June. **Fruit** ½" blue berries, dusted white.

BROADLEAF TREES: opposite compound

SMALL TREES & SHRUBS: alternate simple

AUTUMN OLIVE
Elaeagnus umbellata, Oleaster Family
Bark brown, smooth. **Leaves** alternate simple, 2-3", toothless, wavy edges, pale green, silvery beneath. **Fruit** ¼", pale red berry, speckled white. Invasive, from Asia.

EASTERN REDBUD
Cercis canadensis, Pea Family
Bark dark brown-red, grooved. **Leaves** alternate simple, 2-6", heart-shaped. In autumn, yellow. **Flowers** ½", pink-red, in clusters, appear before leaves, April—May. **Fruit** 4" green-brown seed pods. **Try** looking for the bright pink-purple flowers that appear in spring before the leaves.

WITCH HAZEL
Hamamelis virginiana
Witch Hazel Family
Bark smooth gray with rough patches. **Leaves** alternate simple, 3-6", rounded with wavy edges and uneven base. In autumn, yellow. **Fruit** ½" light brown acorn-like seedpod that pops, flinging seeds up to 20' away in late autumn.

EUROPEAN BUCKTHORN
Rhamnus cathartica, Buckthorn Family
Bark brown, corky, with yellow inner bark. **Leaves** opposite simple (a few might be alternate), 2", rounded, fine-toothed, with *thorns at end of twigs*. **Fruit** ¼", dark blue berries. Invasive, from Europe.

DOGWOODS
Cornus spp., Dogwood Family
Bark smooth to scaly, gray-brown to red. **Leaves** most opposite simple, about 3", curved veins following outer edge of leaf. In autumn, deep maroon. **Flowers** white, 4 petals. **Fruit** blue to white berries on red stems.

HONEYSUCKLES
Lonicera spp., Honeysuckle Family
Bark brown, flaky, on thick woody vine or shrub. **Leaves** opposite simple, rounded, toothless. **Flowers** with 5 petals, fragrant, usually flashy yellow, white, or red. **Fruit** small red, yellow, blue, or black berries, paired at leaf axils. Some species invasive.

STRIPED MAPLE
Acer pensylvanicum, Maple Family
Bark green-brown with vertical white stripes. **Leaves** opposite simple, 2-10", 3 (sometimes 5) pointed lobes, double-toothed. In autumn, yellow. **Fruit** 1" paired seeds with wings.

STAGHORN SUMAC
Rhus hirta (typhina), Cashew Family
Bark dark brown, smooth, horizontal lines. **Leaves** alternate compound, with 11-31 4" leaflets, toothed, on *hairy twigs*. In autumn, scarlet red. **Fruit** 6" red furry spike of berries.

SMALL TREES & SHRUBS: opposite simple

MULTIFLORA ROSE
Rosa multiflora, Rose Family
Bark smooth red-gray, shrubby, with long arching dense *thorny branches*.
Leaves alternate compound, 5", with 7-11 rounded toothed 1½" leaflets.
Flowers 1", white, 5 petals, June—July. **Fruit** red rose hips, leathery, last through winter. Invasive, from Asia.

BLADDERNUT
Staphylea trifolia, Bladdernut Family
Bark gray-green, white streaks.
Leaves opposite compound, 2-6", football-shaped, fine-toothed, usually 3 leaflets. **Fruit** 1-2" papery hanging bladder. **Try** finding a "bladder" from the tree. Shake it to hear the seeds inside.

ELDERBERRY
Sambucus canadensis
Honeysuckle Family
Bark smooth, brown-gray. **Leaves** opposite compound, 4-11", with 4" rounded, toothed leaflets. In autumn, green-yellow. **Flowers** small, white, in flat-topped clusters, June—July. **Fruit** small purple-black berries clustered on branching stems.

SMALL TREES & SHRUBS: compound

A young hiker wraps his arms around a large American Beech in Smith Woods in Trumansburg, New York. This stand of virgin woods contains massive trees, some over 250 years old.

WILDFLOWERS AND OTHER PLANTS

This section covers relatively small flowering plants that die back each year, including common **wildflowers**, **vines**, and **grasses**. A few species are woody, but too small to be considered shrubs. The other species are **herbaceous** (non-woody). Since most wildflowers are identified by the **color of their flowers** to a casual observer, they are organized in this fashion here. Within each color section, the species are in order by the time of their **beginning bloom**. As one becomes more skilled in plant identification, it becomes easier to identify a plant by its leaves when the flower is not in bloom and to recognize different plant families.

Please be sure to always observe or photograph wildflowers without harming them. They are an important part of the ecosystem and some are protected by law.

WHITE FLOWERS

BLOODROOT
Sanguinaria canadensis, Poppy Family
Bloom: April—May
Flower 1½", single, eight to ten petals, golden center, separate stalk from leaf. **Leaves** single, pale, deeply lobed, embracing the flower stalk, butterfly-shaped, *connected in center*, persist after flower has died. **Fruit** green seed capsule. **Found** in woods. 6", taller after bloom.

HEPATICA
Hepatica sp., Buttercup Family
Bloom: April—June
Flower 1", six to ten petal-like sepals, single on hairy stalks, some pink or lavender. **Leaves** three-lobed, rounded (*H. americana*) or pointed (*H. acutiloba*), persist through winter. **Found** in woods and uplands. 5".

TWINLEAF
Jeffersonia diphylla, Barberry Family
Bloom: April—May
Flower 1½", single, eight to ten petals, golden center, separate stalk from leaf. **Leaves** *nearly divided in center*, fan-shaped, persist after flower has died. **Fruit** green capsule. **Found** in woods. 8", taller after bloom.

DUTCHMAN'S BREECHES
Dicentra cucullaria, Poppy Family
Bloom: April—May
Flowers ¾", yellow tipped, waxy, *two spurs* pointing up, dangling, clustered on leafless stalk. **Squirrel Corn** (*D. canadensis*) has no spurs on *heart-shaped* flowers. **Leaves** feathery, dissected. **Fruit** green pod. **Found** in woods. 10".

MAY-APPLE
Podophyllum peltatum, Barberry Family
Bloom: April—June
Flower 1½", single, waxy, nodding, six to nine petals, in fork of paired leafstalks. **Leaves** paired or single (immature plants), umbrella-like, deeply lobed. **Fruit** yellow berry, August. **Found** in forests and openings. 12".

LARGE-FLOWERED TRILLIUM
Trillium grandiflorum, Lily Family
Bloom: April—June
Flower 4", single, three petals with wavy edges, waxy, turning pink. **Leaves** three, large, broad, diamond-shaped. **Fruit** red berries. **Found** in moist woods. 18".

WHITE FLOWERS

FALSE SOLOMON'S SEAL
Smilacina racemosa, Lily Family
Bloom: April—June
Flowers small, clustered in spike at end of plant, fragrant. True **Solomon's Seal** (*Polygonatum biflorum*) has yellow flowers dangling at leaf axils. **Leaves** long, pointed, parallel veined, alternating along curved stem. **Fruit** speckled green to red berries. **Found** in woods. 2'.

CUT-LEAVED TOOTHWORT
Dentaria laciniata, Mustard Family
Bloom: April—June
Flowers ½", four petals, clustered, sometimes pink. **Leaves** in threes, split into three, slender, toothed. **Toothwort** (*D. diphylla*) has wider leaves. **Fruit** seed pods. **Found** in moist woods. 8".

GARLIC MUSTARD
Alliaria officinalis, Mustard Family
Bloom: April—June
Flowers ¼", four petals like other mustards, clustered. **Leaves** *triangular* or heart-shaped, sharp-toothed, with strong garlic odor. **Fruit** long slender seed pods. **Found** on wood edges and trail-sides. 3'. Invasive, from Europe.

WILD STRAWBERRY
Fragaria virginiana, Rose Family
Bloom: April—June
Flowers 1", five round petals, in flat clusters, separate stalk from leaves. **Leaves** in threes, coarse-toothed, oval, on hairy stalks. **Fruit** red, rounded, with *embedded small seeds*. **Found** in fields, open areas. 5".

RED RASPBERRY
Rubus idaeus, Rose Family
Bloom: May—June
Flowers ½", in clusters, five small pedals that fall away. **Leaves** compound with three to seven leaflets, toothed, hairy, on *prickly reddish stems*. **Fruit** ¾" cluster of multiple red drupelets, ripens in late summer. **Found** along trail-sides. Thicket-forming.

WHITE BANEBERRY
Actaea pachypoda, Buttercup Family
Bloom: May—June
Flowers tiny, four to ten petals, small in clustered upright spikes. **Leaves** alternate, compound, toothed. **Fruit** white berries with black dot, on red stalk. **Found** in woods. 24". **Warning** poisonous if ingested.

VIRGINIA WATERLEAF
Hydrophyllum virginianum
Waterleaf Family
Bloom: May—August
Flowers ½", five petals, bell-shaped, five stamens longer than petals, sometimes pale violet. **Leaves** compound, toothed, five to seven lobed, often spotted. **Found** in wet meadows or woods. 24".

WOOD STRAWBERRY
Fragaria vesca, Rose Family
Bloom: May—August
Flowers ¾", five round petals, in flat clusters, separate stalk from leaves. **Leaves** in threes, coarse-toothed, pointed, on hairy stalks. **Fruit** round, red, with small *seeds on surface*. **Found** in rocky woods, pastures. 4". Alien, from Europe.

DAISY FLEABANE
Erigeron annuus, Aster Family
Bloom: May—October
Flowers ¾", forty to seventy *hair-like* rays, some pinkish, around yellow center. **Leaves** narrow, pointed, toothed, on hairy stems. **Found** along trail-sides and fields. 4'.

QUEEN ANNE'S LACE
Daucus carota, Parsley Family
Bloom: May—October
Flowers tiny, in 4" flat cluster, with single tiny purple floret in center, curl up when old. **Leaves** fine, divided. **Found** in trail-sides, fields, dry ground. 4'. **Warning** poisonous, may irritate skin. Alien, from Europe, can be invasive.

WHITE SWEET CLOVER
Melilotus alba, Pea Family
Bloom: May—October
Flowers tiny, in slender tapering 4" clusters. **Leaves** in threes, fragrant, toothed. **Found** in field edges, trail-sides. 8'. Invasive, from Eurasia.

FRAGRANT BEDSTRAW
Galium triflorum, Bedstraw Family
Bloom: June—August
Flowers tiny, four petals, in groups of three, green-white. **Leaves** slender, in whorls of six, on smooth stems. **Fruit** prickly, stick to clothing. **Found** along trail-sides. Mat-forming.

OX-EYE DAISY
Chrysanthemum leucanthemum
Aster Family
Bloom: June–August
Flower 2", single, white rounded rays surrounding yellow center. **Leaves** narrow, many lobes. **Found** in fields and trail-sides. 2'. Alien, from Europe.

YARROW
Achillea millefolium, Aster Family
Bloom: June–August
Flowers small, five rays, in flat tight 3" clusters. **Leaves** fern-like, gray-green, soft, fragrant when crushed. **Found** in fields, trail-sides. 2'. Alien, from Europe.

INDIAN PIPE
Monotropa uniflora, Wintergreen Family
Bloom: June–September
Flower single, nodding bell, four to five petals, waxy, sometimes pink turning black. **Leaves** white, scaly. **Found** in shady woods. 7". Parasitic on mycorrhizal fungi.

BONESET
Eupatorium perfoliatum, Aster Family
Bloom: July–September
Flowers small, fuzzy, in flat clusters. **Leaves** triangular, toothed, opposite, *connected* around hairy stem, veiny, wrinkled. **Found** near watersides. 4'.

WHITE FLOWERS

YELLOW TO ORANGE FLOWERS

POKEWEED
Phytolacca americana
Pokeweed Family
Bloom: July—September
Flowers tiny, in long clusters, green-white petal-like sepals. **Leaves** large, rounded. **Fruit** green to purple-black, clustered on hanging red stalks. **Found** along wood edges, trail-sides. 8'.
Warning poisonous if ingested.

WHITE SNAKEROOT
Eupatorium rugosum, Aster Family
Bloom: July—October
Flowers small, fuzzy, in flat clusters. **Leaves** triangular, toothed, opposite, narrow, long stalks. **Hyssop-Leaved Thoroughwort** (*E. hyssopifolium*) is similar but has slender grass-like leaves. **Found** in moist woods. 4'.

TROUT-LILY
Erythronium americanum, Lily Family
Bloom: April—May
Flower 1", single, six curled petals, often brown on backside, nodding. **Leaves** paired, broad, from base of flower stalk, mottled or spotted. **Found** in woods. 8".

COMMON CINQUEFOIL
Potentilla simplex, Rose Family
Bloom: April—June
Flower ½", five petals, rise from runners, separate stalk from leaves. **Leaves** in *fives*, toothed, rounded. **Found** in fields, dry woods. 12" runners.

50

COMMON DANDELION
Taraxacum officinale, Aster Family
Bloom: April—September
Flower 1-2", single floret with many tiny flowers, bright yellow to orange. **Leaves** jagged, lobed, at base of hollow milky stem. **Fruit** tiny brown seed with cotton-like attachment, carried by wind. **Found** in fields, trail-sides, open areas. 6". Alien, from Europe.

YELLOW LADY'S SLIPPER
Cypripedium calceolus, Orchid Family
Bloom: May—July
Flower 2½", sac-like "slipper" with twisted petals. **Leaves** large, parallel-veined, alternately clasping hairy stem. **Found** in wet woods. 2'.

COMMON BUTTERCUP
Ranunculus acris, Buttercup Family
Bloom: May—August
Flower 1", five overlapping petals. **Leaves** deeply divided, five to seven segments, on hairy stems. **Found** in fields, trail-sides. 2'. Alien, from Europe.

YELLOW WOOD-SORREL
Oxalis sp., Wood-sorrel Family
Bloom: May—September
Flowers ½", five petals, clustered. **Leaves** in threes, heart-shaped, folded, clover-like, on hairy stems. **Fruit** small seed pods. **Found** along trail-sides, fields. 10".

YELLOW TO ORANGE FLOWERS

WILD PARSNIP
Pastinaca sativa, Parsley Family
Bloom: May—October
Flowers tiny, in 2" clusters. **Leaves** compound, toothed, five to fifteen leaflets, on rigid stems. **Found** in fields, trail-sides. 5'. Alien, from Europe.

MONEYWORT
Lysimachia nummularia
Primrose Family
Bloom: June—August
Flowers 1", five petals, paired along trailing vine. **Leaves** paired, opposite, rounded. **Found** on moist ground. Mat-forming. Alien, from Europe.

BIRD'S-FOOT TREFOIL
Lotus corniculatus, Pea Family
Bloom: June—September
Flowers ½", in clusters of three to six. **Leaves** five parts, clover-like. **Fruit** seed pod. **Found** along trail-sides, fields. 15". Alien, from Europe.

COMMON MULLEIN
Verbascum thapsus
Snapdragon Family
Bloom: June—September
Flowers small, clustered on club-like spike. **Leaves** large, rounded, densely hairy, downy soft, numerous along stem. **Found** along trail-sides, open areas. 4'. Alien, from Eurasia.

COMMON ST. JOHNSWORT
Hypericum perforatum
St. Johnswort Family
Bloom: June–September
Flowers 1", five petals, black dots on edges, bushy stamens, clustered. **Leaves** rounded, dotted. **Found** in fields, trail-sides. 20". Alien, from Europe.

BLACK-EYED SUSAN
Rudbeckia hirta, Aster Family
Bloom: June–October
Flower 3", single, ten to twenty rays around *dark brown center*, on slender hairy stem. **Leaves** rounded to slender, hairy, lower leaves toothed. **Found** in fields, trail-sides, open areas. 3'. **Warning** bristly stem and leaves can irritate skin.

JEWELWEED
Impatiens capensis
Touch-me-not Family
Bloom: July–September
Flower 1", hanging, spurred, yellow to orange-red. **Leaves** rounded, slightly toothed, pale below. **Fruit** seed pod that will erupt when touched. **Found** in moist, shady areas. 6".

GOLDENRODS
Solidago spp., Aster Family
Bloom: July–October
Flowers tiny, in flat or branching clusters. **Leaves** long, slender, some toothed or smooth. **Found** in fields, trail-sides, forests. 3-5'. Many different species in area.

YELLOW TO ORANGE FLOWERS

PINK TO RED FLOWERS

JERUSALEM ARTICHOKE
Helianthus tuberosus, Aster Family
Bloom: August—October
Flower 3", several rays around yellow center. **Leaves** thick, *rough*, broad, on *hairy stems*, opposite towards base. **Found** in fields, trail-sides. 6'.

SPRING BEAUTY
Claytonia virginica, Purslane Family
Bloom: April—May
Flowers ¾", five white petals with pink veins. **Leaves** paired midway up stem, slender, smooth. **Found** in moist woods. 10".

WILD GINGER
Asarum canadense, Birthwort Family
Bloom: April—May
Flower 1½", red-brown, three pointed lobes, nodding near ground from fork of paired leaves, cup-shaped, downy, ill-scented. **Leaves** heart-shaped, thick, downy, on hairy stalks. **Found** in woods. 8".

RED TRILLIUM
Trillium erectum, Lily Family
Bloom: April—June
Flower 2½", single, nodding on slender stalk, three maroon petals, ill-scented. **Leaves** three, large, diamond-shaped. **Found** in woods. 12".

WILD COLUMBINE
Aquilegia canadensis, Buttercup Family
Bloom: May–June
Flowers 1½", nodding, lantern-shaped, with five spurs. **Leaves** compound, divided in threes. **Found** in rocky woods. 18".

WILD GERANIUM
Geranium maculatum, Geranium Family
Bloom: May–June
Flowers 1¼", 5 petals, faint pink-purple. **Leaves** deeply divided in 5 sections, hairy, toothed. **Fruit** long "crane's bill" that flings seeds away from plant when dry. **Found** in woods, shady trail-sides. 20".

FRINGED POLYGALA
Polygala paucifolia, Milkwort Family
Bloom: May–June
Flowers 3/4", two pink-purple side wings (sepals), bushy tassel at end of tube-like petals. **Leaves** rounded, evergreen. **Found** in woods. 4".

HEDGE BINDWEED
Convolvulus sepium
Morning-glory Family
Bloom: May–September
Flower 3", trumpet-shaped, pink to white, striped, running along vine. **Leaves** triangular, *squared* on back corners like arrowhead. **Field Bindweed** (*C. arvensis*) has smaller leaves pointed on back corners. **Found** along trail-sides, watersides.

PINK TO RED FLOWERS

PINK TO RED FLOWERS

RED CLOVER
Trifolium pratense, Pea Family
Bloom: May—September
Flowers tiny, clustered in rounded 1" heads. **Leaves** rounded, in threes with pale markings, on hairy stems. **Found** in fields and trail-sides. 20". Alien, from Europe.

HERB ROBERT
Geranium robertianum
Geranium Family
Bloom: May—October
Flowers ½", five petals, usually paired. **Leaves** divided, *fern-like*, three to five segments, on hairy stems. **Found** in rocky woods. 16".

COMMON MILKWEED
Asclepias syriaca, Milkweed Family
Bloom: June—August
Flowers ½", in rounded 2" clusters. **Leaves** large, broad, rounded on downy stems. **Fruit** bumpy seed pod filled with cottony seeds. **Found** in fields, dry areas. 4'.

CROWN VETCH
Coronilla varia, Pea Family
Bloom: June—August
Flowers small, in 1" clover-like rounded clusters. **Leaves** rounded, small, compound, paired, on creeping stems. **Found** along trail-sides. 18". Invasive, from Europe.

DEPTFORD PINK
Dianthus armeria, Pink Family
Bloom: June—August
Flowers ½", five toothed petals, bright pink, speckled with white. **Leaves** slender, grass-like. **Found** in dry fields, trail-sides. 20". Alien, from Europe.

PURPLE-FLOWERING RASPBERRY
Rubus odoratus, Rose Family
Bloom: June—August
Flower 2", five petals. **Leaves** 6", maple-like, toothed, on hairy stems. **Fruit** red berry, cup-shaped. **Found** in rocky woods, ravines. Shrub-like. 6'.

KNAPWEED
Centaurea sp., Aster Family
Bloom: June—September
Flowers 1", feathery, lavender, atop scale-like bracts. **Leaves** slender, some toothed or lobed. **Found** in fields, trail-sides. 3'. Alien, from Europe, some species invasive.

BULL THISTLE
Cirsium vulgare, Aster Family
Bloom: June—September
Flower 2", bright lavender tuft, on spiny flower bracts. **Leaves** pointed, thorny, pale beneath, on prickly stem. **Found** in fields, trail-sides. 5'. Alien, from Europe.

PINK TO RED FLOWERS

BLUE TO VIOLET FLOWERS

SPOTTED JOE-PYE-WEED
Eupatorium maculatum, Aster Family
Bloom: July—September
Flowers small, fuzzy, in flat 5" clusters, pink to purple. **Leaves** narrow, toothed, whorled in fours and fives on deep purple stem. **Found** in moist fields. 4'.

TEASEL
Dipsacus sylvestris, Teasel Family
Bloom: July—September
Flowers tiny, lavender, clustered on rounded 3" head with spines. **Leaves** paired, narrow, toothed, on prickly stem. **Found** in fields, trail-sides. 5'. Alien, from Europe.

COMMON PERIWINKLE
Vinca minor, Dogbane Family
Bloom: April—June
Flower 1", petals lopsided, star in center. **Leaves** evergreen, glossy, paired, on creeping stem. **Found** on forest floors. 6". Invasive, from Europe, spread from gardens.

COMMON BLUE VIOLET
Viola sororia
Violet Family
Bloom: April—June
Flowers ½", five petals, spur on bottom petal, three lower petals veined. **Leaves** heart-shaped, on separate stalks from flowers. **Found** in damp woods. 8".

DAME'S ROCKET
Hesperis matronalis, Mustard Family
Bloom: May—July
Flowers 1", four petals, clustered on top, sometimes pink or white. **Leaves** alternate, toothed. **Fruit** long slender seed pods. **Found** along trail-sides, wood edges. 3'. Alien, from Europe, spread from gardens.

BLUE-EYED GRASS
Sisyrinchium sp., Iris Family
Bloom: May—July
Flowers ½", six petals with points, yellow center. **Leaves** slender, grass-like. **Fruit** round seed pod. **Found** in meadows, marshes, open areas. 12".

COW VETCH
Vicia cracca, Pea Family
Bloom: May—August
Flowers ½", numerous, clustered on one side. **Leaves** compound, eight to twelve pairs of rounded leaflets, on downy stems, vine-like with tendrils. **Found** in fields, trail-sides. 3'. Alien, from Europe.

BITTERSWEET NIGHTSHADE
Solanum dulcamara, Tomato Family
Bloom: May—September
Flowers ½", five petals, yellow anthers, clustered. **Leaves** two lobes at base, on climbing vine. **Fruit** green to red berries. **Found** in moist wood edges. Alien, from Eurasia. **Warning** berries poisonous.

BLUE TO VIOLET FLOWERS

CHICORY
Cichorium intybus, Aster Family
Bloom: June–September
Flowers 1½", light blue fringed rays, stalk-less. **Leaves** toothed at base, few found on stem. **Found** in fields, trail-sides. 3'. Alien, from Europe.

HEAL-ALL
Prunella vulgaris, Mint Family
Bloom: June–September
Flowers 1/4", hooded, clustered on rounded oblong head, violet to pink. **Leaves** rounded, slightly toothed or not, opposite, on four-sided stems. **Found** in fields, trail-sides. 6".

BLUE VERVAIN
Verbena hastata, Vervain Family
Bloom: July–September
Flowers tiny, five petals, clustered on narrow branching spikes. **Leaves** toothed, narrow, on grooved four-sided stem. **Found** along trail-sides, watersides. 6'.

NEW ENGLAND ASTER
Symphyotrichum novae-angliae
Aster Family
Bloom: August–October
Flowers 1", over forty rays, deep violet, golden center, clustered. **Leaves** toothless, crowded, clasping hairy stem. **Found** in fields, moist areas. 5'.

SKUNK CABBAGE
Symplocarpus foetidus, Arum Family
Bloom: February—April
Flowers tiny, clustered on round spadix hidden inside cloak-like purple-brown 6" spathe. **Leaves** large, unfurl after flowers, cabbage-like, foul-odor when crushed. **Found** in wet woods, swamps. 2'.

EARLY MEADOW-RUE
Thalictrum dioicum, Buttercup Family
Bloom: April—May
Flowers 1/4", drooping clustered, yellow anthers, four to five sepals. **Leaves** compound, three to five rounded blunt lobed leaflets. **Found** in woods, ravines. 2'.

BLUE COHOSH
Caulophyllum thalictroides
Barberry Family
Bloom: April—June
Flowers ½", six-pointed, green to yellow. **Leaves** compound, seven to nine leaflets, lobed. **Fruit** deep blue berries. **Found** in woods. 2'.

POISON IVY
Toxicodendron radicans, Cashew Family
Bloom: May—June
Flowers very small, in clusters, sometimes white-yellow. **Leaves** in threes, toothed or lobed, growing alternate on climbing hairy vine, erect, or trailing along ground. **Fruit** white berry. **Found** in woods and trail-sides. **Warning** touching any part can cause severe skin irritation.

GREEN TO BROWN FLOWERS

GREEN TO BROWN FLOWERS

JACK-IN-THE-PULPIT
Arisaema triphyllum, Arum Family
Bloom: May—June
Flower tiny, clustered at base of spadix, wrapped in cup-like spathe with striped flap. **Leaves** large, in threes, single (males) or paired (female). **Fruit** cluster of green to deep red berries. **Found** in woods. 2'.

SMOOTH BROME GRASS
Bromus inermis, Grass Family
Bloom: May—June
Flowers branched in 1" spikes, erect or hanging. **Leaves** 10", blade-like, flat. **Found** in fields, trail-sides. 3'. Alien, from Eurasia, forage crop.

COMMON CATTAIL
Typha latifolia, Cattail Family
Bloom: May—July
Flowers tiny, tightly packed in clusters on top of stem, white (male) above green-brown (female) cylinder. **Leaves** long, stiff, blade-like. **Found** near watersides. 9'.

WILD GRAPE
Vitus sp., Grape Family
Bloom: May—July
Flowers small, green. **Leaves** large, lobed or toothed, on climbing woody vine. **Fruit** purple to black berries. **Found** along trail-sides, wood edges, climbing trees.

TIMOTHY
Phleum pratense, Grass Family
Bloom: June—August
Flowers tiny, in dense 4" cylinders.
Leaves 7", blade-like, flat. **Found** in fields, trail-sides. 3'. Alien, from Europe, hay crop.

VIRGINIA CREEPER
Parthenocissus quinquefolia
Grape Family
Bloom: June—August
Flowers tiny, yellow-green, clustered.
Leaves toothed, in fives, sometimes threes, on climbing or creeping vine, turn scarlet red in autumn. **Fruit** small purple berries. **Warning** berries poisonous.

NETTLES
Urtica sp., Nettle Family
Bloom: June—September
Flowers tiny, in hanging clusters at leaf axils. **Leaves** toothed, heart-shaped, on hollow four-sided stem covered in stinging hairs. **Found** along trail-sides, woods, wet areas. 4'. Alien, from Eurasia. **Warning** stinging acidic hairs cause burning when touched.

YELLOW FOXTAIL
Setaria glauca, Grass Family
Bloom: July—September
Flowers tiny, in dense 4" cylinders, soft and bristly. **Leaves** 10" blade-like, flat, smooth, with rough edges. **Found** in fields, trail-sides. 3'.

GREEN TO BROWN FLOWERS

INSECTS

Insects are members of the Phylum Arthropoda, the largest animal group. **Arthropods** are invertebrates with an **exoskeleton** and segmented bodies and jointed legs. This group includes insects, crustaceans, arachnids, as well as extinct trilobites. Insects are distinguished from other arthropods by having three main body regions — the **head**, **thorax**, and **abdomen** — with three pairs of legs that attach to the thorax. The head has a pair of antennae, compound eyes, and specialized mouthparts. Being the only arthropod capable of powered flight, an insect usually has four wings attached to the thorax. An insect's reproductive and other internal organs are located in the abdomen. Insects go through **metamorphosis** throughout their lives. This transformation is either complete — including an egg, **larva**, **pupa**, and **adult** stage — or it is simple where there is no pupal stage and the insect gradually develops from a **nymph** into an adult. As an insect develops, it must molt, or shed its rigid exoskeleton so that it can continue to grow. Many insects can produce different sounds as well as sense them. These are used for mating and communication. Some have advanced vision and chemical receptors as well.

This section organizes the insects by order and family, so that out of the over 90,000 species of insects in North America, you can find a close relative to the one that you have found. Butterflies and Moths, insects that both belong to the Lepidoptera Order, are located in their own separate section beginning on page 90.

MAYFLIES
Isonychia and other genera
Mayfly Order
Length ¾". **Color** brown to yellow. **Traits** slender abdomen, 4 wings held above body when resting, antennae tiny, 2 or 3 long trailing hair-like tails. **Habitat** near water, eggs laid over water, nymphs aquatic, adults often swarm during mating for day or two and then die. **Food** debris, small aquatic invertebrates.

YELLOW-LEGGED MEADOWHAWK
Sympetrum vicinum
Common Skimmer Dragonfly Family
Length 1¼". **Color** thorax red with yellow legs, wings clear. **Traits** wings held flat out to sides when resting. **Habitat** ponds and fields, eggs inserted into aquatic plants, naiads (nymphs) live in water before becoming adults. **Food** insects.

EBONY JEWELWING
Calopteryx maculata
Broad-winged Damselfly Family
Length 1¾". **Color** metallic green (male) or brown (female), wings black. **Traits** slender abdomen, large eyes, wings held folded back when resting. **Habitat** wooded streams, eggs inserted into aquatic plants, naiads (nymphs) live in water before becoming adults. **Food** small insects.

BLUETS
Enallagma sp.
Narrow-winged Damselfly Family
Length 1¼". **Color** bright blue with black, wings clear. **Traits** slender abdomen, large eyes, wings held folded back when resting. **Habitat** ponds, eggs inserted into aquatic plants, naiads (nymphs) live in water before becoming adults. **Food** small insects.

EUROPEAN EARWIG
Forficula auricularia
Common Earwig Family
Length ½". **Color** shiny brown to black, legs light brown. **Traits** short forewings, long antennae, pincers curved (male) or straight (female) at hind end. **Habitat** damp woods, under logs and leaf litter, nocturnal, eggs laid in burrows in spring and guarded by female. **Food** flowers, fruit, insect larvae.

INSECTS

CHINESE MANTIS
Tenodera aridifolia, Mantid Family
Length 3". **Color** brown to green.
Traits triangular head, large eyes, elongated body, forelegs fold as if praying with spines to grasp prey.
Habitat fields, gardens, eggs laid and overwinter in papery case, Alien, from China. **Food** insects.

NORTHERN WALKINGSTICK
Diapheromera femorata
Common Walkingstick Family
Length 3-4". **Color** brown to green.
Traits legs and body long and stick-like, antennae long, wingless. **Habitat** woods, on trees and shrubs, tiny white eggs with band hatch nymphs in spring. **Food** plant fluids, leaves.

DIFFERENTIAL GRASSHOPPER
Melanoplus differentialis
Short-horned Grasshopper Family
Length 1-2". **Color** yellow, green to brown, glossy. **Traits** large hind legs with stripes, clear wings, short antennae, large eyes. **Habitat** fields, weeds, eggs overwinter in soil. **Food** plants. **Sound** scraping trill.

TRUE KATYDID
Pterophylla camellifolia
Long-horned Grasshopper Family
Length 2". **Color** green. **Traits** long antennae, oval wings, triangular body, males with brown at base of forewings (stridulation field, used for sound production). **Habitat** woods, in trees, eggs overwinter in bark of twigs. **Food** leaves. **Sound** 2 or 3 raspy harsh pulses often in chorus from trees at night beginning in July.

FIELD CRICKET
Gryllus pennsylvanicus, Cricket Family
Length 1". **Color** black to dark brown. **Traits** long antennae, 2 cerci (segmented sensory organs) at rear of abdomen. **Habitat** fields, meadows, woods, eggs or nymphs overwinter in soil. **Food** scavengers, plants, fruit, seeds, insects. **Sound** triple chirps or long trills.

CAMEL CRICKET
Ceuthophilus sp.
Camel and Cave Cricket Family
Length 1½". **Color** brown to dark brown. **Traits** large folded hind legs, curved humpback, long antennae, wingless. **Habitat** damp places, under logs or stones, overwinter as nymphs, eggs laid in soil. **Food** leaf debris.

CICADAS
Magicicada and *Tibicen* spp.
Cicada Family
Length 1-2". **Color** black or green. **Traits** stocky body, large eyes. **Habitat** woods, eggs laid in bark, nymphs burrow in ground later climbing out to molt, molts can be seen clinging to trees or posts (pictured). **Food** tree fluids. **Sound** loud intense buzzing or whining from males.

SCARLET-AND-GREEN LEAFHOPPER
Graphocephala coccinea
Leafhopper Family
Length 3/8". **Color** bright blue-green and red, legs yellow. **Traits** body tiny and tapered, pointed head, short antennae. **Habitat** fields, eggs laid in plant tissues, adults overwinter in leaf litter. **Food** plant fluids.

INSECTS

INSECTS

MEADOW SPITTLEBUG
Philaenus spumarius
Froghopper Family
Length ¼". **Color** yellowish. **Traits** oval body, short wings and antennae, produces foamy froth on plants for protection (pictured). **Habitat** fields, brush, nymph of adult froghoppers, eggs laid on plants in fall. **Food** plant fluids.

WATER BOATMEN
Corixa sp., Water Boatmen Family
Length ½". **Color** brown to black. **Traits** oval body, paddle-like hind legs. **Habitat** ponds, slow streams, eggs laid on aquatic vegetation. **Food** algae, decaying plants.

COMMON WATER STRIDER
Gerris remigis, Water Strider Family
Length ¾". **Color** gray to black. **Traits** long hind and middle legs skate across water, flat slender body. **Habitat** lakes, ponds, streams, eggs laid on water edges, adults overwinter in leaves near water. **Food** small insects.

LEAF-FOOTED BUG
Acanthocephala terminalis
Leaf-footed Bug Family
Length 1½". **Color** gray to black, orange tips on antennae. **Traits** hind legs flat and leaf-like, narrow head, hump back, can give off foul odor if disturbed, walks slowly but can fly. **Habitat** fields, wood edges, tiny green eggs laid on plants. **Food** plant fluids.

SMALL MILKWEED BUG
Lygaeus kalmii
Chinch and Seed Bug Family
Length ¾". **Color** black with orange "X" pattern. **Traits** oval body, 4 or 5 veins on forewing. The **Boxelder Bug** (*Leptocoris trivittatus*) has less orange not in an "X" and many veins on forewing. **Habitat** fields with milkweed, eggs laid on milkweed in spring. **Food** seeds and leaves of milkweed.

GREEN STINK BUG
Acrosternum hilare, Stink Bug Family
Length ¾". **Color** green with yellowish edges, nymphs colorful (pictured). **Traits** flat and shield-like, can give off foul odor if disturbed, walks slowly but can fly. **Habitat** fields, barrel-shaped eggs laid on underside of leaves in rows. **Food** plant fluids.

LOCUST BORER
Megacyllene robiniae
Long Horned Beetle Family
Length 1". **Color** black with yellow stripes, legs reddish. **Traits** long antennae. **Habitat** woods, fields, eggs laid in Black Locust trees in fall, larvae make tunnels in spring. **Food** Black Locust as larvae, Goldenrod pollen as adults.

SIX-SPOTTED GREEN TIGER BEETLE
Cicindela sexguttata
Tiger Beetle Family
Length ¾". **Color** metallic green with 6 white spots. **Traits** fast flyers and runners. **Habitat** open sunny areas, eggs laid in holes in ground, larvae burrow emerging the following summer as adults. **Food** insects, spiders, can bite if handled.

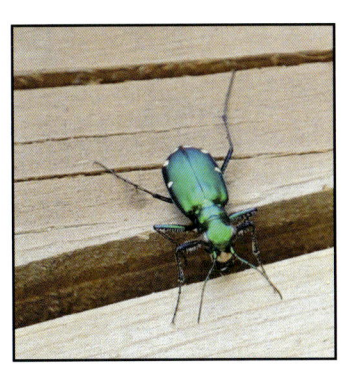

INSECTS

LARGE WHIRLIGIG BEETLES
Dineutus sp., Whirligig Beetle Family
Length 1/2". **Color** glossy black.
Traits flattened ovals, paddle-like hind legs, grooved forewings, swims in circles. **Habitat** ponds, slow streams, lakes, eggs laid on aquatic plants. **Food** small insects.

JAPANESE BEETLE
Popillia japonica, Scarab Beetle Family
Length ½". **Color** metallic copper and green, white tufts on sides. **Traits** oval body, clubbed or fan-like antennae. **Habitat** fields, woods, gardens, eggs laid in soil in summer. Invasive, from Japan. **Food** plant roots as larvae, wide variety of plant leaves as adults.

FIREFLY OR LIGHTNING BUG
Photuris sp., Firefly Family
Length ½". **Color** black-brown with orange and yellow. **Traits** long flat body, males fly at night flashing to females on plants or ground. **Habitat** woods, fields, eggs laid on ground, larvae hibernate, pupate, and emerge following summer. **Food** snails and other small invertebrates as larvae, possibly nectar as adults, sometimes other fireflies.

CONVERGENT LADYBIRD BEETLE
Hippodamia convergens
Ladybird Beetle Family
Length ¼". **Color** red with black spots (usually 13), black with white converging lines on thorax. **Traits** rounded, dome-shaped. **Habitat** fields, woods. **Food** aphids. Sold commercially.

BASSWOOD LEAFMINER
Baliosus nervosus, Leaf Beetle Family
Length ¼". **Color** brown with darker brown markings. **Traits** rectangular body, narrow head, short antennae. **Habitat** woods, eggs laid on leaves, larvae burrow through leaves. **Food** Basswood and other tree leaves.

MOSQUITOES
Aedes and *Culex* spp., Mosquito Family
Length ½". **Color** brown to black. **Traits** slender body, long legs, antennae feathery (males) or thread-like (females), long mouthparts used for sucking. **Habitat** widespread, eggs laid in water, larvae aquatic. **Food** animal blood (females) and plant juices.

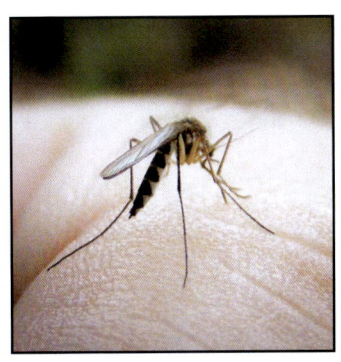

DEER FLY
Chrysops callidus
Horse and Deer Fly Family
Length ½". **Color** black, golden stripes, black patches on wings. **Traits** large eyes patterned with gold or green, antennae spike-like. **Habitat** woods, fields near water. **Food** animal blood (females) and nectar.

CARPENTER ANT
Camponotus pennsylvanicus
Ant Family
Length ½". **Color** black. **Traits** slightly hairy, elbowed antennae, some males and queens with wings, hindwings shorter than forewings, can bite. **Habitat** woods, usually in standing dead trees. **Food** insects, sugars, excavates wood but does not eat it.

INSECTS

EASTERN YELLOW JACKET
Vespula maculifrons, Wasp Family
Length ¾". **Color** bold yellow and black, dark wings. **Traits** stout body, black triangle on abdomen near thorax, will sting if disturbed. **Habitat** woods, fields, parks, nests under debris or underground, builds small light brown paper nest. **Food** scavengers of meat, insects, sugar.

BALD-FACED HORNET
Vespula maculata, Wasp Family
Length ¾". **Color** mainly black with whitish yellow. **Traits** faintly hairy, smokey wings, will sting if disturbed. **Habitat** wood edges, builds football-sized hanging paper nest in trees, fertilized queens overwinter underground or in trees. **Food** nectar, insects.

HONEY BEE
Apis mellifera, Bee Family
Length ½". **Color** gold with black stripes. **Traits** hairy thorax, carries pollen on hind legs, will sting if disturbed. **Habitat** woods, fields, builds wax hive in tree or crevice. Invasive, from Eurasia. **Food** nectar and pollen.

EASTERN BUMBLE BEE
Bombus impatiens, Bee Family
Length ¾". **Color** black and pale yellow. **Traits** densely hairy, can sting but not aggressive. **Habitat** woods, fields, nest underground, queens overwinter underground. **Food** nectar and pollen.

Many insects work hard all spring and summer collecting pollen and nectar from flowering plants throughout this prairie at Cayuga Nature Center. Insects play a large roll in pollinating these plants, allowing them to create seeds and reproduce.

BUTTERFLIES AND MOTHS

Butterflies and Moths are insects that belong to the Order **Lepidoptera**. Butterflies hatch from eggs and begin their larval stage as **caterpillars**. The eggs are laid on a host plant on which the caterpillar feeds during its larval stage. Caterpillars have 6 legs like all insects, but they also have extra pairs of **prolegs** that make them appear worm-like. After feeding, the caterpillar enters the pupal stage and a **chrysalis** is formed. Some butterflies overwinter in this stage. Once emerged from the chrysalis, the adult stage has begun as a butterfly. Most butterflies feed on flower nectar. Mating occurs in the adult stage and the cycle repeats. Moths can usually be distinguished from butterflies by their feather-like antennae, by holding their wings flat and folded back during rest, by duller color, and by activity at night. During the pupal stage, moths pupate in **cocoons** or underground. There are over 180 species of butterflies and moths in Tompkins County.

SILVER-SPOTTED SKIPPER
Epargyreus clarus, Skipper Family
Wingspan 2½". **Color** brown-black, hindwing with silver band on underside. **Traits** lobed hindwings. **Habitat** open woods. **Food** milkweed, red clover, thistles, rarely visits any yellow flowers.

BLACK SWALLOWTAIL
Papilio polyxenes, Swallowtail Family
Wingspan 3-4". **Color** black, yellow spots in rows, orange and black dot on inner edge of hindwing, female with blue on hindwing and less yellow (pictured), male with bolder yellow. **Traits** tails on hindwings. **Habitat** open areas. **Food** red clover, milkweed, thistles.

EASTERN TIGER SWALLOWTAIL
Papilio glaucus, Swallowtail Family
Wingspan 4-6". **Color** yellow with black stripes, female with blue coloring and orange spot on hindwing. **Traits** tails on hindwings. **Habitat** woods, woodland edges. **Food** cherry trees, tulip trees, lilacs.

CABBAGE WHITE
Pieris rapae, White and Sulphur Family
Wingspan 2". **Color** white with black-tipped forewings, 2 black spots on each wing on female, one on male. **Traits** one of the first butterflies to emerge in spring. **Habitat** fields, open areas. **Food** mustards, dandelion, clovers, asters.

ORANGE SULPHER
Colias eurytheme
White and Sulphur Family
Wingspan 2". **Color** yellow-orange with black border, black spot on male. **Habitat** fields, open areas. **Food** dandelion, milkweed, asters, goldenrods.

AMERICAN COPPER
Lycaena phlaeas
Gossamer-wing Family
Wingspan 1". **Color** shiny red-orange with black spots, underside gray-brown. **Traits** striped antennae. **Habitat** open disturbed areas. **Food** buttercup, clover, yarrow, daisy.

BUTTERFLIES

MONARCH
Danaus plexippus, Brushfoot Family
Wingspan 3-5". **Color** bright orange with black veins, white spots, male with 2 small black dots on hindwings. **Traits** white spots on body, migrates in late summer to California and Mexico. **Habitat** fields, marshes, open areas. **Food** milkweed as caterpillar, variety of flowers as adult.

GREAT SPANGLED FRITILLARY
Speyeria cybele, Brushfoot Family
Wingspan 2-4". **Color** tan-orange with black, silver spots on underside of hindwing, female darker. **Habitat** open wet areas. **Food** variety including milkweeds, thistles, vetches, red clover, coneflowers.

MOURNING CLOAK
Nymphalis antiopa, Brushfoot Family
Wingspan 2-4". **Color** black to purple with yellow border, row of blue spots next to border. **Traits** irregular border on wings. **Habitat** woods, open areas, overwinter as adults. **Food** oak tree sap.

RED-SPOTTED PURPLE
Limenitis arthemis astyanax
Brushfoot Family
Wingspan 2-4". **Color** black edged with blue, orange and white spots on tips. **Habitat** woods, overwinter as caterpillars. **Food** sap, carrion, rotting fruit, some viburnum.

VICEROY
Limenitis archippus, Brushfoot Family **Wingspan** 2½-3½". **Color** orange with black veins, white spots, black line across both hindwings making "V" shape. **Traits** mimics Monarch. **Habitat** swamps, wet fields, shrubby areas. **Food** honeydew, carrion, fungi, goldenrods, thistle.

NESSUS SPHINX MOTH
Amphion floridensis Sphinx Moth Family **Wingspan** 2". **Color** red-brown, 2 yellow lines on abdomen. **Traits** stout body, hovers like hummingbird. **Habitat** woods, stream sides, overwinter as caterpillars. **Food** lilac, phlox, herb robert.

EASTERN TENT CATERPILLAR
Malacosoma americanum Tent Caterpillar Moth Family **Wingspan** 2". **Color** brown-white as moth, hairy black with white stripe as caterpillar. **Traits** build nests in forks of branches in spring. **Habitat** near fruit trees. **Food** leaves of fruit trees.

FALL WEBWORM
Hyphantria cunea, Tiger Moth Family **Wingspan** 1". **Color** white with spots as moth, hairy white with black and orange spots as caterpillar. **Traits** build nests at ends of branches in autumn. **Habitat** wood edges. **Food** leaves of hardwood trees.

MOTHS

SPIDERS AND KIN

Spiders are arthropods that are part of the Arachnid Class, which also includes ticks and harvestmen. **Arachnids** have 2 body segments, the **cephalothorax** and the **abdomen**. Located on the cephalothorax are the mouthparts, sensory organs, and limbs. Arachnids generally have 6 pairs of appendages: 4 pairs of limbs, a pair of **palps** used as feelers, and a pair of **chelicerae** used as pincers or fangs. Spiders do not have chewing mouthparts; they use their chelicerae to inject venom to immobilize their prey. Then they regurgitate digestive fluids, and suck up their digested prey. Spiders usually have 8 eyes and no wings or antennae. All spiders produce silk, but only some build webs to trap their prey, which are usually insects. Other spiders hunt down their food by quickness or with the help of camouflage.

BLACK-AND-YELLOW ARGIOPE
Argiope aurantia, Orb Weaver Family
Length ½-1", females larger than males. **Color** abdomen egg-shaped with black and yellow, sometimes orange, legs black with yellow toward body, cephalothorax hairy silver. **Traits** 8 eyes, build large webs. **Habitat** gardens, flowers, shrubs. **Food** insects.

MARBLED ORB-WEAVER
Araneus marmoreus
Orb Weaver Family
Length ¼-½". **Color** abdomen yellow to orange with black markings, legs banded. **Traits** 8 eyes, build large webs. **Habitat** fields, tall grass, shrubs. **Food** insects.

WOLF SPIDERS
Lycosidae, Wolf Spider Family
Length 1". **Color** gray-brown, striped. **Traits** hairy, 8 eyes not in two rows, carries egg sac, young crowd on abdomen after hatched, does not build web. **Habitat** fields, woods, on ground, under rocks. **Food** insects.

GRASS SPIDER
Agelenopsis pennsylvanica
Funnel-web Spider Family
Length ¾". **Color** gray-brown with dark and light stripes. **Traits** long spinnerets at end of abdomen, legs bristly, 8 eyes, weaves funnel-shaped web near ground. **Habitat** grasses. **Food** insects.

NURSERY WEB SPIDER
Pisaurina mira
Nursery Web Spider Family
Length ½". **Color** yellow-brown, band down middle of back. **Traits** narrow, 8 eyes in curved rows, carries egg sac around by mouth, builds tent-like nursery when eggs are about to hatch. **Habitat** fields, woods, tall grasses. **Food** insects.

GOLDENROD CRAB SPIDER
Misumena vatia, Crab Spider Family
Length ¼". **Color** yellow to white with red markings on abdomen. **Traits** legs short flat body, crab-like and held to sides, does not build web. **Habitat** fields, on flowers. **Food** insects.

SPIDERS AND KIN

WOODLOUSE HUNTER
Dysdera crocata, Cell Spider Family
Length ½". **Color** dull red-brown with tan smooth abdomen. **Traits** 6 eyes, large fangs. **Habitat** woods, under logs and rocks. **Food** woodlice.

YELLOW SAC SPIDER
Cheiracanthium inclusum
Prowling Spider Family
Length ⅜". **Color** pale brown-yellow, darkened "feet" and fangs. **Traits** 8 eyes in two rows, long front legs, noticeable spinnerets, faint stripe on abdomen. **Habitat** woods, gardens, homes; weave small sacs for retreat but not to capture prey. **Food** insects, spiders; can bite, moderately venomous to humans.

DADDY-LONG-LEGS OR HARVESTMEN
Opiliones Order
Length ¼". **Color** varies from brown, black, tan, and red, black legs. **Traits** long hair-like legs, 2 body segments joined into one main body segment, 2 eyes. **Habitat** woods, on ground, trees, under logs and rocks. **Food** scavengers, eating small insects and decaying organic matter.

DEER TICK
Ixodes dammini, Hard Tick Family
Length 1/16". **Color** black and brown, reddish abdomen. **Traits** very tiny, wait on long grasses or plants to hitch ride on animal, embed head into skin to suck blood. **Habitat** woods, fields. **Food** animal blood. **Warning** can transmit Lyme Disease.

SPIDERS AND KIN

Not all spiders weave webs to catch prey, but those that do use the spinnerets on their abdomen to produce and string different types of silk. The familiar and noticeable wheel-shaped web above is called an orb web. Other web-building spiders make flat sheet webs, rounded funnel webs, or irregular cobwebs.

LAND SNAILS

Snails are invertebrates that belong to the **Gastropod** Class and to the larger phylum of **mollusks**, which also includes squid, octopus, clams, and mussels. Land snails are typically found in moist woodland habitats. Most build single coiled shells made of calcium carbonate that are used for protection and to retain moisture. **Slugs** are snails with greatly reduced or no shells. A snail moves over surfaces by wave-like contractions of its muscular **foot**. On the front of a snail's head are touch-, light-, and chemical-sensitive tentacles used for navigating and locating food or mates. Near the bottom of the head, inside the mouth, is the **radula**, a ribbon-like structure covered with chitinous teeth, used for scraping and eating food. Most snails feed on decaying organic matter. All of the snails included here are members of the Order **Pulmonata** (air-breathing snails).

- Compiled by Marla Coppolino

DUSKY ARION
Arion subfuscus, Slug Family
Length 2-3". **Description** slug, orange to yellow-brown. **Found** around human habitations and can be a garden pest. Invasive, from Europe.

GRAY FIELDSLUG
Deroceras reticulatum, Slug Family
Length 1-2". **Description** pale brown-gray, emits milky-white mucus when disturbed, can self-amputate its tail when threatened, tail keeled dorsally. **Found** around human habitations and can be a garden pest. Invasive, probably from western Europe.

GLOSSY PILLAR
Cochlicopa lubrica, Pillar Snail Family
Shell length ¼". **Description** shell glossy brown to yellow, translucent, smooth, spindle-shaped. **Found** in and around rotting logs and leaf litter.

DOMED DISC
Discus patulus, Disc Snail Family
Shell width ¼". **Description** shell light brown-gray, spiral shaped, flattened, distinct raised ribs. **Found** sometimes in large numbers but in localized areas.

GRAY-FOOT LANCETOOTH
Haplotrema concavum
Lancetooth Family
Shell width ¾". **Description** shell translucent brown, spiral-shaped, flattened. A predator, elongated neck used to extend into the shells of other snails, specially adapted radula with barbed teeth projections to eat the flesh of other snails. **Found** in woods.

TOOTHED GLOBE
Mesodon zaletus, Liptooth Family
Shell width 1". **Description** shell large, rounded, spiral-shaped, broadly reflected outer lip (outer edge of the opening) and a denticle (a small calcareous projection on the shell surface) on the inner lip surface. **Found** in hardwood forests.

LAND SNAILS

SMALL SPOT
Punctum minutissimum
Spot Snail Family
Shell width less than 1/16", about the size of the period at the end of this sentence, one of the smallest land snails in North America. **Description** shell translucent brown, spiral-shaped, flattened. **Found** in leaf litter.

EUROPEAN AMBERSNAIL
Succinea putris, Ambersnail Family
Shell length 3/8". **Description** shell translucent brown, oval. **Found** near creeks and marshes, amphibious, in large numbers in late summer. Alien, from Europe.

GARLIC GLASS-SNAIL
Oxychilus alliarius, Supercoil Family
Shell width 3/16". **Description** shell translucent, brown, spiral-shaped, flattened, emits a garlic-like odor when disturbed. **Found** in deciduous woods, pine groves, also near human habitations. Alien, from Europe.

BLUE GLASS
Nesovitrea binneyana, Supercoil Family
Shell width 1/8". **Description** shell glossy and translucent, nearly transparent with greenish tinge. **Found** in leaf litter.

LAND SNAILS

Moist leaf litter along forest floors is often the best place to find land snails. Look carefully and closely – many land snails are very small and dull in color. You might also find insects, spiders, centipedes, millipedes, worms, and salamanders living in these leaves.

OTHER FRESHWATER AND LAND INVERTEBRATES

EARTHWORMS
Lambricus and other genera
Earthworm Class
Length 8". **Color** brown-purple. **Traits** moist smooth skin, soft, segmented. **Habitat** moist soil in woods, fields. Some species are native, and some invasive. **Food** decaying organic matter.

ZEBRA MUSSEL
Dreissena polymorpha, Bivalve Class
Length 2" shell. **Color** white-gray to tan with dark stripes. **Traits** triangular shell, attaches to rocks and boats in groups. **Habitat** freshwater rivers and lakes. Invasive, from Europe. **Food** quickly filters small organic material from water.

CRAYFISH
Cambarus, *Orconectes*, and *Procambarus* spp.
Freshwater Crayfish Family
Length 3-5". **Color** brown to red-orange. **Traits** lobster-like, forelimbs with pincers. **Habitat** streams, bays, ponds, some dig burrows. **Food** aquatic plants, invertebrates, fish eggs. 13 species in New York State.

WOODLICE
Armadillidium spp. (Pill Bugs) and *Porcellia* spp. (Sow Bugs) Isopod Order
Length ½". **Color** brown to gray. **Traits** oval with many segmented plates, 7 pairs of legs, 2 antennae, Pill Bugs roll into a ball when disturbed. **Habitat** woods, beneath logs, rocks, leaves. **Food** decaying plant matter, fungus.

MILLIPEDES
Apheloria, *Oxidus*, and *Spirobolus* spp., among others, Millipede Class
Length 1-3". **Color** varied, brown to black, with orange, yellow, red. **Traits** long segmented body, *2 pairs* of legs per body segment, slow, short antennae, roll into ball if disturbed and emit foul odor. **Habitat** woods, fields, under logs and rocks. **Food** decaying plant matter.

CENTIPEDES
Scolopocryptops and *Scutigera* spp., among others, Centipede Class
Length 1-3". **Color** varied, brown to red. **Traits** long segmented body, *one pair* of legs per body segment, long antennae, fast, can bite. **Habitat** woods, under logs and rocks. **Food** insects, spiders.

OTHER INVERTEBRATES

AMPHIBIANS

Amphibians are "cold-blooded" vertebrates (back-boned animals that cannot regulate their body temperature internally) with moist, scaleless skin that is smooth or warty. They have no claws or external ear openings. Their eggs are soft and jelly-like without a shell and are laid in water or moist environments. Most amphibians have a fairly complex life cycle.

Two orders of amphibians live in our area: Caudata (salamanders) and Anura (frogs and toads). **Salamanders** have slender bodies, keep their tails as adults, and most have four limbs that are roughly the same size. Some have vertical grooves along the sides of their body called **costal grooves** that can be used for identification. They feed mainly on invertebrates and often go unnoticed being voiceless and typically living under leaf litter or logs, underground, or in water. **Frogs and toads** have large heads and eyes, two muscular hind legs, and no tail as an adult. Adults feed mostly on invertebrates. Larvae, known as tadpoles, feed mostly on plants. Frogs and toads have excellent hearing and vocal abilities and their calls can often be used to identify them from a distance.

SALAMANDERS

EASTERN NEWT
Notophthalmus viridescens
Newt Family
Length 3-5". **Color** 10-12 *black-ringed red spots*, bright orange as eft stage, olive to red-brown as adult with keeled aquatic tail, yellow belly with black spots. **Traits** no costal grooves, rough skin as eft. **Habitat** woods, streams, ponds. Adults and larvae aquatic, terrestrial during immature eft stage.

SPOTTED SALAMANDER
Ambystoma maculatum
Mole Salamander Family
Length 6-8". **Color** black to gray with irregular rows of large yellow or orange spots. **Traits** stocky body, broad snout, usually 11-13 costal grooves. **Habitat** moist shady woods. Adults terrestrial, larvae aquatic, migrate to vernal ponds to breed in early spring.

BLUE-SPOTTED SALAMANDER
Ambystoma laterale
Mole Salamander Family
Length 4-6". **Color** gray to black, paler belly, blue specks especially on lower sides and limbs. **Traits** flattened tail, 12-14 costal grooves. **Jefferson Salamander** (*A. jeffersonianum*) has a wider snout and longer legs. **Habitat** moist woods. Adults terrestrial, larvae aquatic, migrate to vernal ponds to breed in early spring.

NORTHERN SLIMY SALAMANDER
Plethodon glutinosus
Lungless Salamander Family
Length 4-8". **Color** shiny black with whitish spots, belly gray. **Traits** slimy, 15-17 costal grooves, secretes sticky slime when handled. **Habitat** rocky banks, ravines, shady moist forests. Adults and larvae terrestrial.

NORTHERN 2-LINED SALAMANDER
Eurycea bislineata
Lungless Salamander Family
Length 2-5". **Color** golden yellow to brown band on back, bordered by two black stripes running from each eye to the tail, speckled back, belly yellow. **Traits** slender body, keeled tail, 15-16 costal grooves. **Habitat** rocky creeks, seeps, damp forests. Adults terrestrial, larvae aquatic.

SALAMANDERS

NORTHERN DUSKY SALAMANDER
Desmognathus fuscus
Lungless Salamander Family
Length 2-4". **Color** gray to brown with darker markings. **Traits** line from eye to base of mouth, hind legs larger than front, keeled tail, 13-15 costal grooves. **Mountain Dusky** (*D. ochrophaeus*) has rounded tail. **Habitat** rocky creeks, seeps, springs. Adults terrestrial but usually close to water, larvae aquatic.

NORTHERN SPRING SALAMANDER
Gyrinophilus porphyriticus
Lungless Salamander Family
Length 5-7". **Color** orange-brown to red-yellow, speckled with black. **Traits** flattened nose, keeled tail, 17-19 costal grooves. **Habitat** rocky streams and springs. Adults semiaquatic, larvae aquatic, breed in late autumn and winter.

FOUR-TOED SALAMANDER
Hemidactylium scutatum
Lungless Salamander Family
Length 2-4". **Color** red-brown, white belly with black spots. **Traits** 4 toes each foot (most salamanders have 5 on hind), 13-14 costal grooves, constriction at base of tail. **Habitat** hardwoods, streams, mossy areas. Adults terrestrial, larvae aquatic.

NORTHERN RED-BACKED SALAMANDER
Plethodon cinereus
Lungless Salamander Family
Length 2-4". **Color** gray to black, belly speckled black and white, white specks on sides, *red-backed phase* has reddish stripe from head to tail, *lead-backed phase* all gray to black. **Traits** slender body, 17-22 costal grooves. **Habitat** woods. Adults and larvae terrestrial.

SALAMANDERS

AMERICAN TOAD
Bufo americanus, True Toad Family
Length 2-5". **Color** variable, brown to red, brown spots. **Traits** large parotoid glands behind eyes, warty skin. **Voice** long high-pitched trill. **Habitat** woods to grassy areas.

SPRING PEEPER
Hyla crucifer, Treefrog Family
Length ¾-2". **Color** brown to gray, some rusty orange. **Traits** dark X on back that is sometimes incomplete, large toe pads. **Voice** high short *peep*s. **Habitat** woods near water.

GRAY TREEFROG
Hyla versicolor, Treefrog Family
Length 1-3". **Color** green-brown to gray, dark spots on back, bright yellow-orange under thighs, white below eye, belly white. **Traits** rough skin, large toe pads. **Voice** resonating chirping trills heard from trees. **Habitat** trees or shrubs near water.

BULLFROG
Rana catesbeiana, True Frog Family
Length 3-8". **Color** green-gray to yellow, belly white. **Traits** largest frog species in North America, large eardrum (tympanum) larger on males, *no ridges on back*. **Voice** deep resonating drone. **Habitat** permanent ponds, lakes.

FROGS AND TOADS

GREEN FROG
Rana clamitans, True Frog Family
Length 2-4". **Color** green to brown, belly white with lines or spots, male with bright yellow throat. **Traits** large eardrum (tympanum) larger on males, *ridges on back*. **Voice** loud short *thunks* like a loose banjo string. **Habitat** close to shallow water, water edges, ponds, streams, lakes.

WOOD FROG
Rana sylvatica, True Frog Family
Length 2-3". **Color** dark brown, *black stripe* from nose to behind eye, white on upper lip, bands on legs. **Traits** toes not fully webbed. **Voice** short *quack*s. **Habitat** moist woods, breeds in vernal ponds.

NORTHERN LEOPARD FROG
Rana pipiens, True Frog Family
Length 2-5". **Color** green-brown, dark *circular* spots with light edges, light-colored ridges on back. **Pickerel Frog** (*R. palustris*) has square blotches. **Traits** slender head. **Voice** a creaking snore, clucking. **Habitat** ponds, marshes, open areas, moist vegetation.

This old horse pond at Cayuga Nature Center is now a home for many frogs, as well as fish, birds, insects, mammals, and reptiles. Approach slowly and listen for the calls of different frogs in a chorus around the edges of the pond.

REPTILES

Reptiles are "cold-blooded" (ectothermic) vertebrates that are covered with dry protective scales. They breathe using lungs, breed on land, and lay eggs that are leathery or brittle (some snakes carry eggs internally until their offspring emerge, thus giving birth to live young). Two orders of reptiles live in our area: Squamata (lizards and snakes) and Testudines (turtles and tortoises). **Lizards and snakes** have scales made of keratin that cover a slender body. The outer layer is shed periodically. Lizards have eyelids, external ear openings, and four limbs with clawed feet; snakes lack all of these. **Turtles and tortoises** have an external shell made of bone fused to their backbone and covered with keratin plates called scutes. Some have soft shells. They lack teeth and instead have a beak covered with keratin.

LIZARDS AND SNAKES

NORTHERN COAL SKINK
Eumeces anthracinus, Skink Family
Length 5-7". **Color** brown, black on sides, with four light stripes down back and tail, young have blue tail. **Traits** streamlined, smooth scales. **Habitat** damp woods, stream edges. **Food** insects, spiders.

NORTHERN RINGNECK SNAKE
Diadophis punctatus edwardsii
Colubrid Snake Family
Length 10-30". **Color** black-blue, head dark, *orange band around neck*, belly orange-yellow. **Traits** smooth scales. **Habitat** damp woods or fields. **Food** insects, amphibians, worms.

BLACK RAT SNAKE
Elaphe obsoleta, Colubrid Snake Family
Length 34-101". **Color** black, *throat white*, *belly white-gray* with spots, patterning more visible on young. **Traits** wide head, keeled scales. **Habitat** woods, wood edges, good climber. **Food** birds, rodents.

EASTERN MILK SNAKE
Lampropeltis triangulum triangulum
Colubrid Snake Family
Length 26-52". **Color** gray-tan, brown patches outlined in black, *V-shaped patch* on back of neck. **Traits** smooth scales. **Habitat** open woods, fields. **Food** rodents, birds, snakes.

SMOOTH GREEN SNAKE
Liochlorophis vernalis
Colubrid Snake Family
Length 14-26". **Color** *bright green*, belly white-yellow. **Traits** smooth scales, small, streamlined. **Habitat** moist fields, grassy areas. **Food** insects, spiders.

NORTHERN WATER SNAKE
Nerodia sipedon, Colubrid Snake Family
Length 22-53". **Color** brown-gray, *cross-bands on neck*, dark alternating blotches on body, belly white-gray with spots. **Traits** broad head, ridged scales. **Habitat** freshwater swamps, ponds, streams. *Swimmer.* **Food** fish, frogs, crustaceans, small mammals. **Warning** non-venomous but will strike repeatedly if bothered.

LIZARDS AND SNAKES

LIZARDS AND SNAKES

BROWN SNAKE
Storeria dekayi, Colubrid Snake Family
Length 10-20". **Color** red-brown to gray, *two rows of small dark spots* on back, belly yellow-brown with small dots, dark on top of head. **Red-Belly Snake** (*S. occipitomaculata*) has red belly and three light spots on neck. **Traits** keeled scales. **Habitat** moist woods, marshes. **Food** worms, slugs, snails.

TIMBER RATTLESNAKE
Crotalus horridus, Pit Viper Family
Length 36-70". **Color** *yellow phase* (pictured) yellow-brown with darker brown blotches, black tail; *black phase* darker in color. **Traits** triangular head, thick body, ridged scales, *rattle*. **Habitat** wooded rocky hills, swamps. **Food** birds, amphibians, snakes, small mammals. **Warning** venomous, will rattle and coil in warning.

EASTERN RIBBON SNAKE
Thamnophis sauritus
Colubrid Snake Family
Length 18-40". **Color** black-brown with *three distinct* yellow stripes. **Traits** keeled scales, *slender*. **Habitat** wet fields, marshes, streams, *swims on surface of water*. **Food** frogs, salamanders, fish.

COMMON GARTER SNAKE
Thamnophis sirtalis
Colubrid Snake Family
Length 18-51". **Color** variable, green-blue, pale yellow stripe down center, *checkered or spotted sides*, with yellow line below. **Traits** keeled scales. **Habitat** woods, fields. **Food** frogs, toads, salamanders, worms.

COMMON SNAPPING TURTLE
Chelydra serpentine
Snapping Turtle Family
Length 8-18". **Color** tan to dark brown shell, often algae-covered. **Traits** large head, three *rows of keels on shell* with points on back, long tail. **Habitat** freshwater with muddy bottom. **Food** invertebrates, fish, birds, mammals, plants, carrion. **Warning** has powerful jaws and will bite if disturbed.

PAINTED TURTLE
Chrysemys picta
Box and Pond Turtle Family
Length 4-10". **Color** olive brown shell with *red bars around edges*, red and yellow stripes on head, neck, legs, and tail. **Traits** smooth shell. **Habitat** shallow streams, ponds, lakes with vegetation. **Food** fish, insects, crustaceans, carrion, plants.

SPOTTED TURTLE
Clemmys guttata
Box and Pond Turtle Family
Length 3-5". **Color** black shell with *yellow spots*, spotting on head, neck, and legs. **Traits** smooth shell, small. **Habitat** wet woods, marshes, muddy streams. **Food** worms, crustaceans, insects, amphibian eggs, plants.

WOOD TURTLE
Clemmys insculpta
Box and Pond Turtle Family
Length 5-9". **Color** brown shell, neck and legs red-orange. **Traits** shell rough, wood-like, with growth rings forming *pyramid shapes*, thick tail. **Habitat** cool streams in woods, wet fields. **Food** worms, slugs, insects, tadpoles, fruits.

TURTLES AND TORTOISES

BIRDS

Birds are members of the Class **Aves** and are the only modern animals that have feathers. They are warm-blooded (endothermic, able to regulate body temperature internally), lay eggs, have excellent sight and hearing, and a rigid, lightweight skeleton. There are over 400 species of resident and migratory birds in New York. This section includes 51 species divided into 7 categories: waterfowl, herons and shorebirds, raptors, owls, woodpeckers, song birds, and others.

WATERFOWL

CANADA GOOSE
Branta canadensis, Waterfowl Family
Length 40". **Wingspan** 50". **Color** dark brown-gray, black head and neck with white "chinstrap." **Habitat** open areas, near water; nests on grassy or marshy ground. **Food** grasses, aquatic plants. **Season** March-April, October-November. **Voice** honking.

WOOD DUCK
Aix sponsa, Waterfowl Family
Length 19". **Wingspan** 30". **Color** dark purple back and chest, sides yellow, head green with drooping crest. **Habitat** ponds, rivers, swamps; nests in cavities. **Food** acorns, seeds. **Season** March-November. **Voice** high whistle, squeal when frightened.

MALLARD
Anas platyrhynchos, Waterfowl Family
Length 23". **Wingspan** 35". **Color** gray, head glossy green. **Habitat** shallow waters, ponds, rivers; nests on ground near water. **Food** seeds in water. **Season** year-round. **Voice** female makes well-known *quack* sound, male rasping quack or whistle.

GREAT BLUE HERON
Ardea herodias, Heron Family
Length 46". **Wingspan** 72". **Color** gray, dark crown. **Habitat** near streams, ponds, meadows, uplands; nests in dead trees. **Food** fish, small mammals, frogs, other prey. **Season** April-November. **Voice** hoarse squawk.

GREEN HERON
Butorides virescens, Heron Family
Length 18". **Wingspan** 26". **Color** dark gray-green, brown neck. **Habitat** streams, ponds; nests in trees. **Food** fish, invertebrates. **Season** April-October. **Voice** harsh *skyeew*.

KILLDEER
Charadrius vociferus, Plover Family
Length 11". **Wingspan** 24". **Color** brown, chest white with two black bands. **Habitat** open areas, fields; nests in gravel. **Food** invertebrates. **Season** March-October. **Voice** drawn out *kill-deer*.

AMERICAN WOODCOCK
Scolopax minor, Sandpiper Family
Length 11". **Wingspan** 18". **Color** pale gray-black, speckled. **Habitat** woods, fields; nests on ground; often active at dusk. **Food** worms, insects. **Season** March-November. **Voice** nasal *peeent*.

HERONS & SHOREBIRDS

TURKEY VULTURE
Cathartes aura
American Vulture Family
Length 26". **Wingspan** 67". **Color** brown-black, head bare red. **Habitat** over woods, fields, open areas; nests in logs and crevices. **Food** carrion. **Season** April-October. **Voice** hiss or cluck, rarely heard.

SHARP-SHINNED HAWK
Accipiter striatus
Hawk and Eagle Family
Length 11". **Wingspan** 23". **Color** gray, rusty breast, square tail. **Habitat** forests, shrubby areas; nests in trees. **Food** small birds. **Season** year-round. **Voice** high *kiw kiw kiw*.

RED-TAILED HAWK
Buteo jamaicensis
Hawk and Eagle Family
Length 19". **Wingspan** 49". **Color** dark brown, breast white with black streaks below, tail pale orange below. **Habitat** fields, woodland edges; nests in tall trees. **Food** small mammals. **Season** year-round. **Voice** rasping *keeerrr*.

AMERICAN KESTREL
Falco sparverius, Falcon Family
Length 9". **Wingspan** 22". **Color** orange, blue-gray, black bands on white face. **Habitat** meadows, fields; nests in cavities. **Food** insects, small mammals. **Season** year-round. **Voice** high *killy killy*.

RAPTORS

BARRED OWL
Strix varia, Owl Family
Length 21". **Wingspan** 42". **Color** brown-gray, spotted and streaked. **Habitat** woods, swamps; nests in cavities or uses crow or hawk nests; nocturnal. **Food** small mammals. **Season** year-round. **Voice** hooting, *who who hoo-hoo*.

GREAT HORNED OWL
Bubo virginianus, Owl Family
Length 22". **Wingspan** 44". **Color** gray-brown, striped, ear tufts on head. **Habitat** woods, forest edges, fields; uses abandoned nests; nocturnal, but seen at dusk and day. **Food** medium-sized mammals. **Season** year-round. **Voice** rhythmic hooting.

EASTERN SCREECH-OWL
Otus asio, Owl Family
Length 9". **Wingspan** 20". **Color** red to gray, striped, ear tufts on head. **Habitat** forest edges, open woods; nests in cavities; nocturnal, roosting in cavities during day. **Food** insects, rodents. **Season** year-round. **Voice** whistled trill, horse-like whinny.

RED-BELLIED WOODPECKER
Melanerpes carolinus
Woodpecker Family
Length 9". **Wingspan** 16". **Color** barred black, chest white with faint red, head red on top and back. **Habitat** woods, feeders; nests in cavities of dead trees or fence posts. **Food** fruits, seeds, insects. **Season** year-round. **Voice** rolling *quirrr*.

DOWNY WOODPECKER
Picoides pubescens, Woodpecker Family
Length 7". **Wingspan** 12". **Color** black and white, male with red patch on back of head, beak shorter than Hairy. **Habitat** woods; nests in cavities of dead parts of trees. **Food** insects, fruit, seeds, suet. **Season** year-round. **Voice** rapid whinny, soft *pik*.

HAIRY WOODPECKER
Picoides villosus, Woodpecker Family
Length 9". **Wingspan** 15". **Color** black and white, male with red patch on back of head, beak longer than Downy. **Habitat** woods; nests in cavities of dead parts of trees. **Food** insects, fruit, seeds, suet. **Season** year-round. **Voice** high *peek*.

PILEATED WOODPECKER
Drycopus pileatus, Woodpecker Family
Length 17". **Wingspan** 29". **Color** black, head with red crest, white on neck and face. **Habitat** hardwood forests; nests in cavities of dead trees. **Food** insects, fruits, nuts. **Season** year-round. **Voice** single or series of loud *wek* notes.

TREE SWALLOW
Tachycineta bicolor, Swallow Family
Length 6". **Wingspan** 15". **Color** shiny blue, breast white, notched tail. **Habitat** fields, water; nests in cavities. **Food** berries, insects. **Season** March-May, August-October. **Voice** series of whistles, or *cheets*.

BARN SWALLOW
Hirundo rustica, Swallow Family
Length 7". **Wingspan** 15". **Color** blue-black, breast light orange, notched tail. **Habitat** fields, ponds; usually nests on man-made structures. **Food** insects. **Season** April-September. **Voice** squeaky twittering, or *vit-vits*.

BLUE JAY
Cyanocitta cristata
Crow and Jay Family
Length 11". **Wingspan** 16". **Color** bright blue, chest white, blue crest on head, black stripe on neck. **Habitat** woods; nests in trees. **Food** insects, seeds, acorns. **Season** year-round. **Voice** harsh *jay jay*, or *queed-le*.

AMERICAN CROW
Corvus brachyrhynchos
Crow and Jay Family
Length 18". **Wingspan** 39". **Color** all black. **Habitat** open areas, woods, fields; nests in trees. **Food** insects, worms, fruit, nuts, small animals, carrion, other songbird eggs. **Season** year-round. **Voice** hoarse *caw*.

WHITE-BREASTED NUTHATCH
Sitta carolinensis, Nuthatch Family
Length 6". **Wingspan** 11". **Color** gray-blue, face and breast white, black crown. **Habitat** woods, will walk up and down tree trunks; nests in natural cavities or woodpecker holes. **Food** insects, seeds. **Season** year-round. **Voice** nasal *wer-wer-wer*, or *reenk*.

SONG BIRDS

SONG BIRDS

TUFTED TITMOUSE
Baeolophus bicolor, Chickadee Family
Length 6½". **Wingspan** 9¾". **Color** pale gray, orange flanks. **Habitat** deciduous woods, feeders; nests in natural cavities or woodpecker holes. **Food** insects and seeds. **Season** year-round. **Song** *peter-peter-peter*, whistled.

BLACK-CAPPED CHICKADEE
Poecile atricapillus, Chickadee Family
Length 5". **Wingspan** 8". **Color** gray, chin and cap black, breast white. **Habitat** woods; nests in boxes, natural cavities, or woodpecker holes. **Food** insects, spiders, seeds. **Season** year-round. **Voice** pleasant descending *fee-beee*, or raspy *chicka-dee-dee-dee*.

HOUSE WREN
Troglodytes aedon, Wren Family
Length 5". **Wingspan** 6". **Color** brown-gray, lightly striped. **Habitat** wood edges, shrubs; nests in boxes, natural cavities, or woodpecker holes; aggressive. **Food** insects. **Season** April-October. **Voice** nasal scolding, trills, rattles.

EASTERN BLUEBIRD
Sialia sialis, Thrush Family
Length 7". **Wingspan** 13". **Color** bright blue, breast orange. **Habitat** open fields; nests in boxes and cavities. **Food** insects, fruit. **Season** year-round. **Voice** pleasant soft *cheer-cheery* whistles, chatter.

VEERY
Catharus fuscescens, Thrush Family
Length 7". **Wingspan** 12". **Color** brown-red, breast white with faint spots. **Habitat** woods, swamps; nests on or near ground. **Food** insects. **Season** May-September. **Voice** spiraling *veer veer veer*.

WOOD THRUSH
Hylocichla mustelina, Thrush Family
Length 8". **Wingspan** 13". **Color** brown-red, breast white with black spots. **Habitat** shaded forests; nests low in trees or shrubs. **Food** invertebrates, insects, fruit. **Season** May-September. **Voice** pleasant, flute-like *eee-oo-ay?*.

AMERICAN ROBIN
Turdus migratorius, Thrush Family
Length 10". **Wingspan** 17". **Color** dark gray, rusty orange breast. **Habitat** woods; nests in trees or human structures. **Food** worms, berries. **Season** March-November, some year-round. **Song** whistled repeated *cheer-e-o cheer-e-me*.

GRAY CATBIRD
Dumetella carolinensis Mockingbird Family
Length 9". **Wingspan** 11". **Color** gray, with darker crown. **Habitat** woods, shrubs; nests in shrubs. **Food** insects, fruit. **Season** May-October. **Voice** mimics, cat-like *mee-ooww*, or rambling scrambled warble.

SONG BIRDS

SONG BIRDS

BROWN THRASHER
Taxostoma rufum, Mockingbird Family
Length 12". **Wingspan** 13". **Color** brown-red, breast white with black streaks. **Habitat** woods, thickets, fields; nests on ground or in shrubs. **Food** insects, invertebrates. **Season** April-October. **Voice** mimics, phrases repeated twice.

NORTHERN MOCKINGBIRD
Mimus polyglottos, Mockingbird Family
Length 10". **Wingspan** 14". **Color** gray, breast pale gray. **Habitat** shrubby fields; nests in shrubs or trees. **Food** insects, fruit. **Season** year-round. **Voice** mimics, 2-6 repeated phrases.

EUROPEAN STARLING
Sturnus vulgaris, Starling Family
Length 9". **Wingspan** 16". **Color** glossy black-green, speckled. **Habitat** fields, buildings; nests in cavities. Invasive, from Europe. **Food** seeds, fruit, insects, invertebrates. **Season** year-round. **Voice** mixed chatter, whistles, chuckles.

YELLOW WARBLER
Dendroica petechia
Wood Warbler Family
Length 5". **Wingspan** 8". **Color** yellow-brown, breast bright yellow, male with reddish streaks on chest. **Habitat** wet areas, shrubs; nests in shrubs or trees. **Food** insects, some fruit. **Season** April-September. **Voice** high, sharp, rapid, *sweet sweet sweet I'm so sweet*.

OVENBIRD
Seiurus aurocapillus
Wood Warbler Family
Length 6". **Wingspan** 10". **Color** brown, breast white with black streaks, two stripes on crown with orange center. **Habitat** shaded forests; nests on ground. **Food** insects. **Season** May-September. **Voice** harsh *tea-cher tea-cher tea-CHER*, increasing in volume.

SCARLET TANAGER
Piranga olivacea, Tanager Family
Length 7". **Wingspan** 12". **Color** red, wings and tail black, female yellow. **Habitat** forests, tops of trees; nests in trees, well away from trunk. **Food** insects. **Season** May-October. **Voice** hoarse robin-like whistles, or descending *chik-burr*.

NORTHERN CARDINAL
Cardinalis cardinalis, Grosbeak Family
Length 9". **Wingspan** 12". **Color** red, crest on head, black face, female brown-red. **Habitat** wood edges, shrubs, feeders; nests in shrubs or small trees. **Food** seeds, fruit, insects. **Season** year-round. **Voice** many variations of whistling, *beeww beeww beeww, pitchew pitchew*.

SONG SPARROW
Melospiza melodia
American Sparrow Family
Length 6". **Wingspan** 8". **Color** brown-red, striped, breast white with brown center spot and brown streaks. **Habitat** open areas, shrubs, near water; nests on ground or in shrubs. **Food** seeds, grass, berries, insects. **Season** year-round. **Voice** complex trills, whistles, *swee swee swee too-ee tri-tri-tri-tri*.

SONG BIRDS

DARK-EYED JUNCO
Junco hyemalis
American Sparrow Family
Length 6". **Wingspan** 9". **Color** gray, belly white. **Habitat** woods, shrubs; nests on ground, sometimes on branches. **Food** seeds, insects, on ground. **Season** year-round. **Voice** short trill.

HOUSE SPARROW
Passer domesticus
Old World Sparrow Family
Length 6". **Wingspan** 10". **Color** dark brown and gray, throat and upper chest black. **Habitat** fields, buildings; nests in cavities. Invasive, from Europe. **Food** seeds, insects. **Season** year-round. **Voice** series of chirps, or husky *chir-up*.

RED-WINGED BLACKBIRD
Agelaius phoeniceus, Blackbird Family
Length 9". **Wingspan** 13". **Color** black, shoulder with red and yellow patch. **Habitat** wet, marshy areas, fields; nests near ground in marshy areas. **Food** seeds, invertebrates. **Season** March-November. **Voice** gurgling, trilling *konk-o-reeee*.

BROWN-HEADED COWBIRD
Molothrus ater, Blackbird Family
Length 7". **Wingspan** 12". **Color** glossy black, head brown. **Habitat** open woods; female lays eggs in native songbirds' nests. **Food** seeds, invertebrates. **Season** April-November. **Voice** gurgles and whistles.

COMMON GRACKLE
Quiscalus quiscula, Blackbird Family **Length** 13". **Wingspan** 17". **Color** all black, iridescent, eyes yellow. **Habitat** open woods, fields, watersides; nests in trees. **Food** seeds, invertebrates. **Season** March-November, some year-round. **Voice** unpleasant gurgling *k-sheek*.

BALTIMORE ORIOLE
Icterus galbula, Blackbird Family **Length** 9". **Wingspan** 12". **Color** black, breast and shoulder orange, white on wings. **Habitat** open woods; nests in trees in hanging nest. **Food** caterpillars, fruit, nectar. **Season** May-September. **Voice** rich whistling, 4-8 phrases.

HOUSE FINCH
Carpodacus mexicanus, Finch Family **Length** 6". **Wingspan** 10". **Color** brown, forehead and chest red. **Habitat** woods, shrubby areas, feeders; nests in trees or on buildings. Invasive, from western United States. **Food** seeds, grains, fruits. **Season** year-round. **Voice** varied warbling, beginning high and ending in low *veeer*.

AMERICAN GOLDFINCH
Carduelis tristis, Finch Family **Length** 5". **Wingspan** 9". **Color** yellow, crown and wings black. **Habitat** fields, wood edges; nests in shrubs in open areas. **Food** tree buds, seeds, insects. **Season** year-round. **Voice** sweet rising and falling *to-WEE to-WEE to-WEE tweer tweer ti ti ti,* or *pur-chick-or-ee*.

OTHERS

WILD TURKEY
Meleagris gallopavo, Partridge Family
Length 46". **Wingspan** 64" (female smaller). **Color** dark brown, iridescent, bare head. **Habitat** open woodlands; nests on ground; roosts in trees at night. **Food** nuts, seeds, insects, salamanders. **Season** year-round. **Voice** many variations of clucks, chirps, gobbles.

MOURNING DOVE
Zenaida macroura
Pigeon and Dove Family
Length 12". **Wingspan** 18". **Color** light brown-gray. **Habitat** open woodlands, fields, shrubby areas; nests in dense trees or on human structures. **Food** seeds, fruit, insects. **Season** year-round. **Voice** mournful *coo-o coo coo coo*.

RUBY-THROATED HUMMINGBIRD
Archilochus colubris
Hummingbird Family
Length 4". **Wingspan** 5". **Color** shiny green, chest white, red throat. **Habitat** woods, forest edges; usually nests near water. **Food** flower nectar, insects. **Season** May-September. **Voice** soft *chips*.

CHIMNEY SWIFT
Chaetura pelagica, Swift Family
Length 5". **Wingspan** 14". **Color** dark gray. **Habitat** fields, towns; nests in chimneys or hollows. **Food** insects. **Season** April-October. **Voice** hard rapid *chips*.

Many bird houses can be seen throughout Cayuga Nature Center along the woods and in the prairie. These provide nesting locations for birds that nest in cavities, such as the Eastern Bluebird. It is important to monitor these boxes to make sure that the aggressive House Sparrow does not force out our native birds.

MAMMALS

Mammals are warm-blooded (endothermic) animals that are nearly all born live rather than from eggs. They have **mammary glands** that produce milk for their young and skin glands used for many functions. Most are covered in hair or fur to aid in maintaining a constant body temperature. Seven orders of Mammals (not including humans) live in the area: Opossums, Shrews and Moles, Bats, Hares and Rabbits, Rodents, Carnivores, and Hoofed Mammals. Most mammals have advanced senses of sight, hearing, and smell and so are often difficult to encounter in the wild. For this reason, signs of each mammal's presence have been included in their description, such as tracks, burrows, scat, or sounds. The mammals here are **organized by overall size**, from smallest to largest. Length measurements are of the mammal's head and body, not including the tail.

DEER MOUSE
Peromyscus maniculatus
Mouse and Rat Family
Length 2½-4". **Color** gray to red-brown, belly and feet white, tail bicolored. **Signs** hind tracks ⅝" with 5 spread toes, 4 toes on front, pile of small chewed nuts, secondary user of 1½" wide burrow openings, nests in any shelter available. **Habitat** woods, fields, nocturnal. **Food** nuts, acorns, seeds, insects.

LITTLE BROWN BAT
Myotis lucifugus, Plainnose Bat Family
Length 2½-4". **Wingspan** 9". **Color** brown, glossy, belly pale white. **Signs** erratic flight, audible squeaks. **Habitat** woods, near water, nocturnal, migrate to caves in winter. **Food** insects. **Big Brown Bat** (*Eptesicus fuscus*) is larger, paler brown, flies straight and fast. There are many other bats in this area.

NORTHERN SHORT-TAILED SHREW
Blarina brevicauda, Shrew Family
Length 3-4". **Color** dark gray, tail dark hairless. **Signs** tracks ¼" narrow with 5 toes on front and hind, dragging tail, 1" wide burrow openings. **Habitat** woods, fields, marshes. **Food** insects, snails, worms, other invertebrates, mice. Poisonous saliva.

MEADOW VOLE
Microtus pennsylvanicus
Mouse and Rat Family
Length 3½-5". **Color** gray to brown, belly silver, tail bicolored and short. **Signs** hind tracks ½" with 5 spread toes, 4 toes on front, grass cuttings, 1½" wide burrow openings. **Habitat** moist areas near streams or lakes, woods, fields. **Food** grasses, seeds, bark, insects.

STARNOSE MOLE
Condylura cristata, Mole Family
Length 4½-5". **Color** dark brown to gray-black, nose pink with fleshy tentacles. **Signs** 12" wide mound of soil with 2" wide burrow openings. **Habitat** moist areas, near streams or lakes, swims well. **Food** worms, insects.

EASTERN CHIPMUNK
Tamias striatus, Squirrel Family
Length 5-6". **Color** red-brown, white stripe on sides edged in black, belly white. **Signs** hind tracks under 2" with 5 toes, 4 toes on front, clean 2" wide burrow openings near woods, sharp *chip* or *chuck-chuck-chuck* sound. **Habitat** woods, brush. **Food** seeds, nuts, bulbs, fruit, insects, eggs.

MAMMALS

MAMMALS

FLYING SQUIRRELS
Glaucomys spp., Squirrel Family
Length 5½-6½". **Color** olive-brown, belly white, loose skin on sides from front to hind legs, northern species (*G. sabrinus*, larger with gray-white belly) and southern species (*G. volans*) in area. **Signs** high-pitched twittering sound, eyeshine red-orange. **Habitat** woods, nocturnal. **Food** seeds, nuts, eggs, insects, some meat.

RED SQUIRREL
Tamiasciurus hudsonicus
Squirrel Family
Length 7-8". **Color** red-brown, belly white. **Signs** hind tracks 1" with 5 toes, 4 toes on front, piles of eaten nuts, noisy angry ratcheting chatter. **Habitat** woods, swamps. **Food** seeds, nuts, fungi, eggs.

EASTERN GRAY SQUIRREL
Sciurus carolinensis, Squirrel Family
Length 8-10". **Color** gray, belly lighter, tail bushy and darker, also a black colored phase. **Signs** hind tracks 2½" with 5 toes, 4 toes on front, nests in 12-18" nest of leaves and twigs at tops of deciduous trees or in tree holes, chattering and clucking sounds. **Habitat** woods. **Food** nuts, seeds, fungus, fruits, bark.

LONG-TAILED WEASEL
Mustela frenata, Weasel Family
Length 8-10½". **Color** brown, belly yellow-white, tail brown with black tip, some white in winter, **Short-tailed Weasel** (*M. erminea*) is smaller. **Signs** tracks 1" with 5 toes (usually only 4 seen), cache of dead rodents. **Habitat** near water, swims, climbs. **Food** small mammals, birds.

COMMON MUSKRAT
Ondatra zibethicus, Mouse & Rat Family **Length** 10-14". **Color** dense brown, glossy, belly silver, tail black and hairless. **Signs** hind tracks 2" partially webbed, front 1", 5 toes, dragging tail, 2-3' rounded houses of marsh vegetation, 5-6" wide burrow openings, in banks. **Habitat** marshes, ponds, lakes, streams. **Food** aquatic plants, clams, frogs, fish.

MINK
Mustela vison, Weasel Family **Length** 12-17". **Color** dark brown to black, chin white. **Signs** tracks rounded 1½" with 5 toes (sometimes only 4 seen), 4" wide burrow entrance in bank, eyeshine yellow-green. **Habitat** near streams and lakes, nocturnal, swims. **Food** small mammals, frogs, eggs, fish.

EASTERN COTTONTAIL
Sylvilagus floridanus Hare and Rabbit Family **Length** 14-17". **Color** brown-gray, cotton-like white tail, feet white. **Signs** hind tracks 3" in front of small circular front with 4 toes, nests in depression in ground, secondary user of 5-6" burrows, scat pea-sized pellets. **Habitat** woods, open areas, swamp edges. **Food** vegetation, bark, twigs.

STRIPED SKUNK
Mephitis mephitis, Weasel Family **Length** 13-18". **Color** black with white stripes. **Signs** hind tracks 1½", front 1", both with 5 toes, holes dug for grubs, foul scent, eyeshine dark orange, dens in 5-6" wide burrow openings. **Habitat** woods, prairie, open areas. **Food** mice, eggs, insects, grubs, berries, carrion. **Warning** can spray scent up to 15' if disturbed.

MAMMALS

MAMMALS

VIRGINIA OPOSSUM
Didelphis virginiana, Opossum Family
Length 15-20". **Color** gray to black with longer white hairs, face white, tail hairless pink with black base. **Signs** hind tracks 2" wide, hand-like, opposable thumb, eyeshine dull orange. **Habitat** woods along streams, farms, nocturnal. **Food** insects, carrion, fruits, grains.

WOODCHUCK (GROUNDHOG)
Marmota monax, Squirrel Family
Length 16-20". **Color** light to dark brown-gray, dark feet. **Signs** hind tracks 2" with 5 toes, 4 toes on front, 6-10" wide burrow openings. **Habitat** open woods. **Food** leaves, flowers, grasses.

PORCUPINE
Erethizon dorsatum
New World Porcupine Family
Length 18-22". **Color** black fur with longer white hairs and quills. **Signs** tracks 3" turned inward, squeal and grunt noises, piles of quills, bark removed towards top of tree, eyeshine deep red. **Habitat** woods. **Food** buds, bark, twigs.

RACCOON
Procyon lotor, Raccoon Family
Length 18-28". **Color** gray-brown, black around eyes, rings on tail. **Signs** hind tracks 4" long, 5 toes, front hand-like, dens in hollow trees or holes with 5-6" wide opening. **Habitat** woods, streams, lakes. **Food** fruits, nuts, grains, insects, frogs, bird eggs.

AMERICAN BEAVER
Castor canadensis, Beaver Family
Length 25-30". **Color** dark brown, scaly hairless flat tail. **Signs** hind tracks 5" with 5 webbed toes, stick and mud dams and lodges, tooth marks on stumps, slaps tail on water as warning. **Habitat** streams, lakes, ponds, usually nocturnal. **Food** bark, twigs.

GRAY FOX
Urocyon cinereoargenteus, Dog Family
Length 20-32". **Color** silver-gray, lower sides and legs orange, belly white, black-tipped tail held straight out when running. **Signs** tracks 1½" with 4 clawed toes, dens in hollowed logs or burrows. **Habitat** open woods, brush, climbs trees. **Food** rodents, hares, birds, eggs, insects, berries.

RED FOX
Vulpes vulpes, Dog Family
Length 18-35". **Color** red-orange, legs and feet black, belly white, white-tipped tail held straight out when running. **Signs** front tracks 2½" with 4 furry clawed toes, short yaps, strong scent, 9-12" wide burrows with larger excavated opening. **Habitat** woods, open areas. **Food** rodents, hares, birds, insects, berries.

RIVER OTTER
Lontra canadensis, Weasel Family
Length 25-40". **Color** dark brown to black, glossy fur, belly lighter. **Signs** tracks 3" wide, 5 partially webbed toes, dens in logs or banks with underwater entrances. **Habitat** near water, rivers, lakes. **Food** amphibians, crayfish, fish, turtles.

MAMMALS

MAMMALS

BOBCAT
Lynx rufus, Cat Family
Length 25-40". **Color** orange-brown to gray, barred with black, tail black at top of tip. **Signs** tracks rounded 2" with 4 clawless toes, dens in crevices and hollowed logs. **Habitat** woods, swamps. **Food** small mammals, birds, carrion.

COYOTE
Canis latrans, Dog Family
Length 30-40". **Color** gray to brown-red, belly and throat white, black-tipped tail held down when running. **Signs** front tracks 2 3/8" in straight line with 4 clawed toes, *yaps* or howls, eyeshine green-gold, 1-2' wide entrance hole to den, scat cylindrical full of hair. **Habitat** open woods, prairies. **Food** small mammals, birds, snakes, frogs.

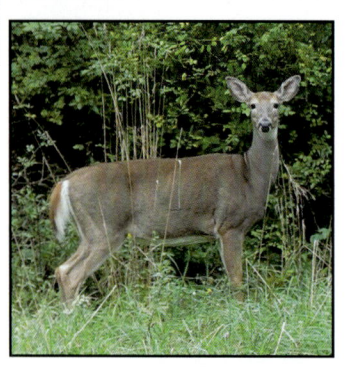

WHITE-TAILED DEER
Odocoileus virginianus, Deer Family
Length 50-80". **Color** brown-red to gray, belly white, white tail up when running, fawns spotted, antlers on males. **Signs** tracks 2-3" two-toed hoof, eaten vegetation, rubbed bark, deer trails, scat small dark pellets, whistling snorts or grunts. **Habitat** woods, swamps, open areas. **Food** twigs, shrubs, acorns, grass, fungi.

BLACK BEAR
Ursus americanus, Bear Family
Length 50-78". **Color** black to dark brown. **Signs** hind tracks 7" long with 5 toes, dens beneath downed trees, boulders, or stumps with 3' wide opening. **Habitat** woods, swamps. **Food** berries, nuts, honey, insects, fish, small mammals, carrion. **Warning** do not feed or approach, can become aggressive.

118

Winter is a great time to search for mammal tracks in the snow. The tracks of an animal can tell you many things, such as the type of animal, its size and speed, the direction and destination of its travel, or its behavior. Tracks can also be found during other times of the year in muddy areas after rain.

Taughannock Falls

PHOTO CREDITS & ACKNOWLEDGMENTS

All photographs by James Dake unless otherwise indicated as follows: Dayna Jorgenson (cover image; two-lined salamander); Jonathan Hendricks (Tully Limestone; joints); Community Science Institute, Ithaca, New York (Cayuga Lake Watershed map); Robert Wesley (Rough-Stalked Feather Moss); Environmental Protection Agency (Virginia Waterleaf); J. Delanoy (White Sweet Clover); Daniel Layton (Yellow Lady Slipper); Marla Coppolino (all land snail photos); Bruce Marlin (Deer Fly); WesDigital (Earwig); Benny Mazur (Mayfly); Tim McCormack (Waterstrider); Kim Fleming (Water Boatman); Tim Gage (Carpenter Ant); Neil Kelley (Convergent Lady Beetle); Jacob Enos (Large Whirligig Beetle); James Jordan (Mosquito); Michael Apel (American Copper); Megan McCarty (Black Swallowtail, Silver-spotted Skipper); Stephen Patrick (Cabbage White); Tommy C. (Orange Sulpher); Michelle Tribe (Crayfish); Hardin MD University of Iowa (Deer Tick); Luc Viatour (Goldenrod Crab Spider); Teresa Yoder (Eastern Newt, Spotted Salamander); Patrick Coin (Slimy Salamander); John Wilson (Northern Spring Salamander); Sander van der Molen (Spring Peeper); John Swanlund (Brown Snake); L.P. (Coal Skink); Per Verdonk (Eastern Milk Snake); Christine Schmidt (Eastern Garter Snake); D.R. (Green Snake); C.F.K. (Northern Watersnake); John Mosesso (Ribbon Snake, Spotted Turtle); Cody Hough (Ringneck Snake); Tim Vickers (Timber Rattlesnake); Wilfried Burns (Wood Turtle); Mike Baird (American Kestrel, Great Blue Heron, Turkey Vulture); Ronald S. (American Woodcock); James Phelps (Barn Swallow, Muskrat, Gray Fox); Robert Barber (Barred Owl); Pierre Bonenfant (Blue Jay); John Benson (Downy Woodpecker, Mourning Dove); Wolfgang Wander (Eastern Screech Owl); Brendan Lally (Great Horned Owl); Manjith Kainickara (Green Heron, American Crow, Common Grackle); Alex Starr (Hairy Woodpecker); Tim Lenz (Killdeer); Eric Baetscher (Mallard); Dominic Sherony (Pileated Woodpecker, Ovenbird, Scarlet Tanager); Chris Breeze (Red-bellied Woodpecker); Christine Fournier (Red-tailed Hawk); Derek Bakken (Sharp-shinned Hawk); Kevin Cole (Wood Duck, House Finch); Ryan Somma (Baltimore Oriole); Will Sweet (American Goldfinch, Black-capped Chickadee); Trisha Shears (Brown Thrasher, European Starling); Jim McCulloch (Chimney Swift); Randen Pederson (Dark-eyed Junco); Andy (Eastern Bluebird); Noel Zia Lee (House Sparrow); Dario Sanches (House Wren); Paul Stein (Northern Cardinal, Bald-faced Hornet); Jeff Kubina (Northern Mockingbird); Jennifer Rensel (Ruby-throated Hummingbird); Gary Irwin (Tufted Titmouse); Bob Lewis (Veery, Yellow Warbler); Matt MacGillivray (White-breasted Nuthatch); Jeff Hoppes (Wood Thrush); Steve Heresy (American Beaver, Black Rat Snake); Sage Ross (Eastern Cottontail); Grendel Khan (Eastern Gray Squirrel); Ken Thomas (Flying Squirrel, River Otter); Erin Silversmith (Groundhog); Gilles Gonthier (Meadow Vole, Northern Short-tailed Shrew); Kenneth Catania (Starnose Mole); Cody Pope (Opossum); Sharon Mollerus (Porcupine); J.M. Wests (Raccoon); Patriot Plaistow John (Red Squirrel); Eric Schmuttenmaer (Striped Skunk); Katie LaSalle-Lowery (Long-tailed Weasel); PeupleLoup (Black Bear, Red Fox); Malcolm (Bobcat); Christopher Bruno (Coyote); Ken (Little Brown Bat); Brendan Lally (Mink).

A special thanks to those who made this book possible: Triad Foundation, Paleontological Research Institution, Cayuga Nature Center, Paula Mikkelsen, Warren Allmon, Rob Ross, Samantha Sands, Brian Gollands, Richard Kissel, Marla Coppolino, Tom Trencansky, Jes Steele, Catherine McCarthy, Marvin Pritts, Jim Spear, Paul Miller, Teresa Yoder, Katie McGlashen, Brian VanPatten, Nat Cleavitt, Linda Spielman, Carolyn Klass, John Wiessinger, Dale Bryner, Charles Eldermire, Katie Bagnall-Newman, and my family and friends for their love and support.

Cayuga Lake at dusk

SOURCES OF MORE INFORMATION

Local Sources

Alden, Peter. 1999. *National Audubon Society Field Guide to the Mid-Atlantic States*. Knopf, New York, 448 pp.

Allmon, Warren D., and Robert M. Ross. 2008. *Ithaca is Gorges: A Guide to the Geology of the Ithaca Area, 4th ed.* Paleontological Research Institution, Ithaca, New York, 28 pp.

Allmon, Warren D., Marvin P. Pritts, Peter L. Marks, Blake P. Epstein, David A. Bullis, and Kurt A. Jordan. 2017. *Smith Woods: The Environmental History of an Old Growth Forest Remnant in Central New York State*. Paleontological Research Institution, Ithaca, New York, Special Publication no. 52, 208 pp.

Ansley, J. E. 2000. *The Teacher-Friendly Guide to the Geology of the Northeastern U.S.* Paleontological Research Institution, Ithaca, New York, Special Publication no. 24, 182 pp.

Bloom, Arthur L. 2018. *Gorges History: Landscapes and Geology of the Finger Lakes Region*. Paleontological Research Institution, Ithaca, New York, Special Publication no. 55, 214 pp.

Bosanko, Dave. 2008. *Fish of New York.* Adventure Publications, Cambridge, Minnesota, 192 pp.

Chapman, William K., ed. *Wildflowers of New York in Color*. Syracuse University Press, Syracuse, New York, 168 pp.

Clausen, R. T. 1949. *Checklist of the Vascular Plants of the Cayuga Quadrangle 42°-43° N, 76°-77° W.* Cornell University Agricultural Experiment Station, Ithaca, New York, Memoir no. 291, 87 pp.

Dieckmann, Jane Marsh. 1986. *A Short History of Tompkins County*. DeWitt Historical Society of Tompkins County, Ithaca, New York, 229 pp.

Donaldson, Alfred Lee. 1921. *A History of the Adirondacks*. Century Company, New York, volume 1. Online at http://books.google.com/books/download/A_History_of_the_Adirondacks.pdf?id=BMWU5iASa4wC&output=pdf&sig=ACfU3U0beXBm8gyEY6LTDaVxxCEpVm|XIA.

Figiel, R. 1995. *Culture in a Glass: Reflections on the Rich Heritage of Finger Lakes Wine*. Silver Thread Books, Lodi, New York, 53 pp.

Gibbs, James P. 2007. *The Amphibians and Reptiles of New York State: Identification, Natural History, and Conservation*. Oxford University Press, New York, 504 pp.

Haine, Peggy. 2002. *Cornell Plantations Path Guide*. Cornell University, Ithaca, New York, 172 pp.

Hinds, James, and Patricia Hinds. 2008. *The Macrolichens of New England*. New York Botanical Garden Press, Bronx, New York, 608 pp.

Isachsen, Y. W., E. Landing, J. M. Lauber, L. V. Rickard, and W.B. Rogers, eds. 2000. *Geology of New York: A Simplified Account*, 2nd ed. New York State Museum Cultural Education Center and New York State Geological Survey, Albany, New York, 300 pp.

Lassoie, James P. 1996. *Forest Trees of the Northeast*. Cornell Cooperative Extension, Ithaca, New York, 227 pp.

Leonard, Mortimer Demarest. 1928. *A List of the Insects of New York with a List of the Spiders and Certain Other Allied Groups*. Cornell University Agricultural Experiment Station, Ithaca, New York, Memoir no. 101, 1091 pp.

Luther, D. D. 1910. Geology of the Auburn-Genoa Quadrangles. *New York State Museum Bulletin* 137, 36 pp.

Magee, Dennis W., and Harry E. Ahles. 2007. *Flora of the Northeast. A Manual of the Vascular Flora of New England and Adjacent New York*, 2nd ed. University of Massachusetts Press, Amherst, 1214 pp.

Mohler, Charles L., Peter Marks, and Sana Gardescu. 2006. *Guide to the Plant Communities of the Central Finger Lakes Region*. New York State Agricultural Experiment Station, Geneva, 128 pp.

Tekiela, Stan. 2005. *Birds of New York*, 2nd ed. Adventure Publications, Cambridge, Minnesort, 324 pp.

Tekiela, Stan. 2006. *Trees of New York*. Adventure Publications, Cambridge, Minnesota, 260 pp.

Wiegand, K. M., and A. J. Eames. 1926. *The Flora of the Cayuga Lake Basin, New York: Vascular Plants*. Cornell University Agricultural Experiment Station, Ithaca, New York, Memoir no. 92, 491 pp.

Wesley, F. R., S. Gardescu, and P. L. Marks. 2008. *Vascular Plant Species of the Cayuga Region of New York State*. Unpublished report, Cornell University, Ithaca, New York, online at http://ecommons.library.cornell.edu/bitstream/1813/9413/6/Wesley_et_al_2008_EXCEL2.pdf.

OTHER GENERAL SOURCES

Bessette, Alan E., Arleen R. Bessette, and David W. Fischer. 1997. *Mushrooms of Northeastern North America*. Syracuse University Press, Syracuse, New York, 582 pp.

Brodo, Irwin M., Sylvia Duran Sharnoff, and Stephen Sharnoff. 2001. *Lichens of North America*. Yale University Press, New Haven, Connecticut, 828 pp.

Brown, Lauren. 1992. *Grasses: An Identification Guide*. Houghton Mifflin Harcourt, Boston, Massachusetts, 256 pp.

Burch, John Bayard. 1962. *How to Know the Eastern Land Snails: Pictured-Keys for Determining the Land Snails of the United States Occurring East of the Rocky Mountain Divide*. WCB/McGraw-Hill, Dubuque, Iowa, 214 pp.

Clemants, Steven, and Carol Gracie. 2005. *Wildflowers in the Field and Forest: A Field Guide to the Northeastern United States*. Oxford University Press, New York, 480 pp.

Comstock, Anna Botsford. 1986. *Handbook of Nature Study* (reprint). Comstock Publishing/Cornell University Press, Ithaca, New York, 887 pp.

Conant, Roger, and Joseph T. Collins. 1998. *A Field Guide to Reptiles & Amphibians of Eastern & Central North America, 4th ed.* Houghton Mifflin Harcort Boston, Massachusetts, Peterson Field Guide no. 12, 616 pp.

Crum, H. A. 1983. *Mosses of the Great Lakes Forest, 3rd ed*. University Herbarium, University of Michigan, Ann Arbor, 417 pp.

Howell, W. Mike, and Ronald L. Jenkins. *Spiders of the Eastern United States: A Photographic Guide*. Pearson Education, Boston, Massachusetts, 363 pp.

Ley, L. & J. Crowe. 1999. *The Enthusiasts Guide to Liverworts and Hornworts of Ontario*. Lakehead University, Thunder Bay, Ontario, Canada, 135 pp.

Newcomb, Lawrence. 1989. *Newcomb's Wildflower Guide*. Little, Brown and Company, Boston, Massachusets, 490 pp.

Ogden, Eugene C. 1981. *Field Guide to Northeastern Ferns.* New York State Museum, Albany, New York, 128 pp.

Opler, Paul A. 1998.. *A Field Guide to Eastern Butterflies*. Houghton Mifflin Harcourt, Boston, Massachusetts, Peterson Field Guide no. 4, 512 pp.

Palmer, E. Laurence, Seymour H. Fowler, & H. Seymour Fowler. 1975. *Fieldbook of Natural History, 2nd ed.* McGraw-Hill, New York, 779 pp.

Peterson, Roger Tory. 1998. *A Field Guide to Wildflowers: Northeastern and North-Central North America.* Houghton Mifflin Harcourt, Boston, Massachusetts, Peterson Field Guide no. 17, 448 pp.

Petrides, George A. 1998. *A Field Guide to Eastern Trees, Peterson Field Guides, 2nd ed.* Houghton Mifflin Harcourt, Boston, Massachusetts, 448 pp.

Reid, Fiona. 2006. *Peterson Field Guide to Mammals of North America, 4th ed.* Houghton Mifflin Harcourt, Boston, Massachusetts, Peterson Field Guide no. 5, 608 pp.

Sibley, David Allen. 2003. *The Sibley Field Guide to Birds of Eastern North America*. Knopf, New York, 432 pp.

WEBSITES

Animal Diversity Web, http://animaldiversity.ummz.umich.edu/site/index.html

Butterflies and Moths of North America, https://www.butterfliesandmoths.org/

Cornell Lab of Ornithology, All About Birds, https://www.allaboutbirds.org/

Finger Lakes Native Plant Society, http://www.fingerlakesnativeplantsociety.org

Insect Identification for the Casual Observer, http://www.insectidentification.org

Mushroom Expert, http://www.mushroomexpert.com

New York Waterfalls and Nature, http://nyfalls.com

Paleontological Research Institution, including Museum of the Earth, Cayuga Nature Center, and Smith Woods: http://www.priweb.org

Tompkins County Flora Project, http://www.plantsystematics.org/

USDA Plants Database, http://plants.usda.gov

Vanderbilt University, Department of Biological Sciences, http://www.cas.vanderbilt.edu/bioimages

Wildflowers of the Northeastern and Northcentral U.S.A., http://www.dclunie.com/eshelton/wildflow/wildind.html

Buttermilk Falls

INDEX

Acanthocephala terminalis 68
Accipiter striatus 100
Acer
    Negundo 38
    pensylvanicum 41
    platanoides 37
    rubrum 37
    saccharum 37
Achillea millefolium 49
Acrosternum hilare 69
Actaea pachypoda 47
Aedes spp. 71
Agelaius phoeniceus 108
Agelenopsis pennsylvanica 79
Aix sponsa 98
Alien Species 7, 8, 47-49, 51-53, 56-60, 62, 63, 66, 84
Alliaria officinalis 46
Ambersnail, European 84
Ambystoma
    Jeffersonianum 89
    laterale 89
    maculatum 89
American Sparrow Family 107, 108
American Vulture Family 100
Amphibians 88-93
Amphion floridensis 77
Anas platyrhynchos 98
Angiosperms 30, 32-42
Annona Family 39
Anura 88
Ant, Carpenter 71
Ant Family 71
Apheloria spp. 87
Apis mellifera 72
Apple 35
Aquilegia canadensis 55
Arachnids 78
Araneus marmoreus 78
Archilochus colubris 110
Ardea herodias 99
Argiope, Black-and-Yellow 78
Argiope aurantia 78
Arion, Dusky 82
Arion subfuscus 82
Arisaema triphyllum 62
Armadillidium spp. 87
Arthropods 20, 64-73, 78-80
Artichoke, Jerusalem 54
Arum Family 61
Asarum canadense 54
Asclepias syriaca 56

Ash
    Green 39
    White 39
Asimina triloba 39
Aspen
    Bigtooth 36
    Quaking 37
Aster, New England 60
Aster Family 40, 48-51, 53, 54, 57, 58, 60
Autumn Olive 40
Aves 98
Baeolophus bicolor 104
Baliosus nervosus 71
Baneberry, White 47
Barberry Family 45, 61
Basswood, American 34
Bat
    Big Brown 112
    Little Brown 112
Bear, Black 118
Bear Family 118
Beaver, American 117
Bedstraw, Fragrant 48
Bedstraw Family 48
Bee
    Eastern Bumble 72
    Honey 72
Bee Family 72
Beech, American 32
Beech Family 32, 33
Beetle
    Convergent Ladybird 70
    Japanese 70
    Large Whirligig 70
    Six-spotted Green Tiger 69
Betula lenta 33
Betula papyrifera 33
Bignonia Family 37
Bindweed
    Field 55
    Hedge 55
Birch
    Black (Sweet) 27, 33
    White (Paper) 33
Birch Family 33, 34
Birds 98-111
Birthwort Family 54
Bivalve Class 86
Blackbird, Red-winged 108
Blackbird Family 108, 109
Bladdernut 42
Blarina brevicauda 113

Bloodroot   44
Blueberry, High-bush   39
Bluebird, Eastern   104
Bluets   65
Bobcat   118
Bombus impatiens   72
Boneset   49
Borer, Locust   69
Box and Pond Turtle Family   97
Box Elder   38
Brachiopods   20
Brachythecium rutabulum   29
Branta canadensis   98
British Soldiers   26
Broadleaf Trees   30, 32-39
Bromus inermis   62
Brushfoot Family   76, 77
Bryozoans   21
Bubo virginianus   101
Buckthorn, European   40
Buckthorn Family   40
Bufo americanus   91
Bug
   Green Stink   69
   Leaf-footed   68
   Lightning   70
   Pill   87
   Small Milkweed   69
   Sow   87
Bullfrog   91
Buteo jamaicensis   100
Butorides virescens   99
Buttercup, Common   51
Buttercup Family   44, 47, 51, 55, 61
Butterflies   74-77
Cabbage, Skunk   61
Calopteryx maculata   65
Camponotus pennsylvanicus   71
Canis latrans   118
Cantharellus cinnabarinus   25
Cardinal, Northern   107
Cardinalis cardinalis   107
Carduelis tristis   109
Carpinus caroliniana   34
Carpodacus mexicanus   109
Carya ovata   38
Caulophyllum thalictroides   61
Cashew Family   41, 61
Castor canadensis   117
Cat Family   118
Catalpa, Northern   37
Catalpa speciosa   37
Catbird, Gray   105
Cathartes aura   100
Catharus fuscescens   105
Cattail, Common   62

Cattail Family   62
Caudata   88
Cayuga Nature Center   23, 73, 93, 111, 121, 125, 136
Cedar
   Eastern Red   30
   White   30
Cell Spider Family   80
Centaurea sp.   57
Centipedes   87
Cephalopods   22
Cercis canadensis   40
Ceuthophilus   67
Chaetura pelagica   110
Chanterelle Family   25
Charadrius vociferus   99
Cheiracanthium inclusum   80
Chelydra serpentine   97
Cherry, Black   36
Chickadee, Black-capped   104
Chickadee Family   104
Chicory   60
Chinch and Seed Bug Family   69
Chipmunk, Eastern   113
Chrysanthemum leucanthemum .   65
Chrysemys picta   97
Chrysops callidus   71
Cicada Family   67
Cicadas   67
Cichorium intybus   60
Cicindela sexguttata   69
Cinnabar Chanterelle   25
Cinquefoil, Common   50
Cirsium vulgare   57
Cladonia cristatella   26
Clams   21, 82
Claytonia virginica   54
Clemmys
   guttata   97
   insculpta   97
Clover,
   Red   56
   White Sweet   48
Cnidaria   21
Coal Skink, Northern   94
Cochlicopa lubrica   83
Cohosh, Blue   61
Colias eurytheme   75
Colubrid Snake Family   94-96
Columbine, Wild   55
Condylura cristata   113
Conifers   30, 31
Convolvulus
   arvensis   55
   sepium   55
Copper, American   75

Corals
    Horn 21
    Rugose 21
    Tabulate 21
Corixa sp. 68
Cornell University 5
Cornus sp. 41
Coronilla varia 56
Corvus brachyrhynchos 103
Cottontail, Eastern 115
Cottonwood, Eastern 36
Cowbird, Brown-headed 108
Coyote 118
Crab Spider Family 79
Crayfish 86
Creeper, Virginia 63
Cricket
    Camel 67
    Field 67
Cricket Family 67
Crinoids 21
Crotalus horridus 96
Crow, American 103
Crow and Jay Family 103
Crustacean Class 86
Cucumber Magnolia 35
Culex spp. 71
Cyanocitta cristata 103
Cypress Family 30
Cypripedium calceolus 51
Daddy-long-legs 80
Daisy, Ox-eye 49
Dame's Rocket 59
Damselfly 65
Danaus plexippus 76
Dandelion, Common 51
Daucus carota 48
Deer, White-tailed 118
Dendroica petechia 106
Dentaria
    diphylla 46
    laciniata 46
Deroceras reticulatum 82
Desmognathus
    fuscus 90
    ochrophaeus 90
Devonian 12-14, 17, 20, 23
Diadophis punctatus edwardsii 94
Dianthus armeria 57
Diapheromera femorata 66
Dicentra cucullaria 45
Didelphis virginiana 116
Dineutus sp. 70
Dipsacus sylvestris 58
Disc Snail Family 83
Discus patulus 83

Dog Family 117, 118
Dogbane Family 58
Dogwood Family 7
Domed Disc 83
Dove, Mourning 110
Dragonfly 65
Dreissena polymorpha 86
Drumlins 16
Drycopus pileatus 102
Dryopteris carthusiana 29
Duck
    Mallard 98
    Wood 98
Dumetella carolinensis 105
Dutchman's Breeches 45
Dysdera crocata 80
Earthworms 86
Earwig, European 65
Earwig Family 65
Echinoderms 21
Ectotherms 94
Elaeagnus umbellata 40
Elaphe obsoleta 95
Elderberry 42
Elm, Slippery 34
Elm Family 34
Enallagma sp. 65
Epargyreus clarus 74
Eptesicus fuscus 112
Erethizon dorsatum 116
Erigeron annuus 48
Erythronium americanum 50
Eumeces anthracinus 94
Eupatorium
    hyssopifolium 50
    maculatum 55
    perfoliatum 49
    rugosum 50
Eurycea bislineata 89
Fagus grandifolia 32
Falco sparverius 100
Falcon Family 100
Feather Moss 29
Fern
    Christmas 29
    New York 28
    Sensitive 28
    Spinulose Wood 29
Finch, House 109
Finch Family 109
Firefly 70
Firefly Family 70
Flavoparmelia caperata 26
Fleabane, Daisy 48
Fly, Deer 71
Forficula auricularia 65

Fox
  Gray   117
  Red   117
Foxtail, Yellow   63
Fragaria
  vesca   47
  virginiana   46
Fraxinus
  americana   39
  pennsylvanica   39
Fritillary, Great Spangled   76
Froghopper Family   68
Frog Family   91, 92
Frogs
  Green   92
  Northern Leopard   92
  Pickerel   92
  Wood   92
Fungi   24-26
Funnel-web Spider Family   79
Galium triflorum   48
Gastropods   22
Geranium, Wild   55
Geranium Family   55, 56
Geranium
  maculatum   55
  robertianum   56
Gerris remigis   68
Ginger, Wild   54
Glass, Blue   84
Glass-snail, Garlic   84
Glaucomys sp.   114
Gleditsia triacanthos   38
Globe, Toothed   83
Goldenrods   53
Goldfinch, American   109
Goose, Canada   98
Gossamer-wing Family   75
Grackle, Common   109
Grape, Wild   62
Grape Family   62, 63
Graphocephala coccinea   67
Grass
  Blue-eyed   59
  Smooth Brome   62
Grass Family   62, 63
Grasshopper
  Differential   66
  Long-horned Family   66
  Short-horned Family   66
Gray Urn Fungus   26
Greenshield, Common   26
Groundhog   116
Grosbeak Family   107
Gryllus pennsylvanicus   67
Gymnosperms   30, 31

Gyrinophilus porphyriticus   90
Hamamelis virginiana   40
Haplotrema concavum   83
Hard Tick Family   80
Hare and Rabbit Family   115
Harvestmen   80
Hawk
  Red-tailed   100
  Sharp-shinned   100
Hawk and Eagle Family   100
Heal-all   60
Heath Family   39
Helianthus tuberosus   54
Hemidactylium scutatum   90
Hemlock, Eastern   32
Hepatica   44
Hepatica
  americana   44
  acutiloba   44
Herb Robert   56
Heron
  Great Blue   99
  Green   99
Heron Family   99
Hesperis matronalis   59
Hickory, Shagbark   38
Hippodamia convergens   70
Hirundo rustica   103
Honeysuckle Family   41, 42
Hornbeam
  Eastern Hop   34
  American   34
Hornet, Bald-faced   72
Horse and Deer Fly Family   71
Hummingbird, Ruby-throated   110
Hummingbird Family   110
Hydrophyllum virginianum   47
Hygrocybe conica   25
Hyla
  crucifer   91
  versicolor   91
Hylocichla mustelina   105
Hypericum perforatum   53
Hyphantria cunea   77
Icterus galbula   109
Impatiens capensis   53
Indian Pipe   49
Insects   65-73
Introduced Species   30, 31, 35, 36
Invasive Species   40-42, 46, 48, 56-58, 70, 72, 82, 86, 106, 108, 109
Invertebrates   64-87
Iris Family   59
Ironwood   34
Isonychia sp.   64
Isopod Order   87

Ivy, Poison   38
Ixodes dammini   80
Jack-in-the-Pulpit   62
Jay, Blue   103
Jeffersonia diphylla   45
Jelly Fungi Class   25
Jewelweed   53
Jewelwing, Ebony   65
Joe-pye-weed, Spotted   58
Juglans nigra   38
Junco, Dark-eyed   108
Junco hyemalis   108
Juniperus virginiana   30
Katydid, True   66
Kestrel, American   100
Killdeer   99
Knapweed   57
Lady's Slipper, Yellow   51
Ladybird Beetle Family   70
Lake, Cayuga   7, 11, 15, 18, 136
Laetiporus sulphureus   24
Lambricus sp.   86
Lampropeltis triangulum triangulum   95
Lancetooth, Gray-foot   83
Landsnails   82-84
Large Cup Fungi Family   26
Leaf Beetle Family   71
Leaf-hopper, Scarlet-and-Green   67
Leaf-hopper Family   67
Leafminer, Basswood   71
Lepidoptera   74
Leucobryum glaucum   29
Lichens   26
Lily Family   45, 46, 50, 54
Limenitis
    archippus   77
    arthemis astyanax   76
Limestone, Tully   22
Linden Family   34
Liochlorophis vernalis   95
Liptooth Family   83
Liriodendron tulipifera   35
Lizards   94
Locust
    Black   38
    Honey   38
Long-horned Beetle Family   69
Lonicera sp.   41
Lotus corniculatus   52
Lungless Salamander Family   89, 90
Lycaena phlaeas   75
Lycoperdon pyriforme   25
Lycosidae   79
Lygaeus kalmii   68
Lynx rufus   118

Lysimachia nummularia   52
Magicicada sp.   67
Magnolia acuminata   35
Magnolia Family   35
Malacosoma americanum   77
Mallard   98
Malus pumila   35
Mammals   112-118
Mantid Family   66
Mantis, Chinese   66
Maple
    Box Elder   38
    Norway   37
    Red   37
    Sugar   37
    Striped   41
Maple Family   37, 38, 41
Marmota monax   116
Marsh Fern Family   28
Mastodons   14
May-apple   45
Mayflies   64
Mayfly Order   64
Meadow-rue, Early   61
Meadowhawk, Yellow-legged   65
Megacyllene robiniae   69
Melanerpes carolinus   101
Melanoplus differentialis   66
Meleagris gallopavo   110
Melilotus alba   48
Melospiza melodia   107
Mesodon zaletus   83
Microtus pennsylvanicus   113
Milkweed, Common   56
Milkweed Family   56
Milkwort Family   55
Millipedes   87
Mimus polyglottos   106
Mink   115
Mint Family   60
Mephitis mephitis   115
Misumena vatia   79
Mockingbird, Northern   105
Mockingbird Family   105-106
Mole, Starnose   113
Mole Family   113
Mole Salamander Family   89
Molothrus ater   108
Monarch   76
Moneywort   52
Monotropa uniflora   49
Moraines   16
Morning-glory Family   55
Mosquitoes   71

Moss
  Feather 29
  White Cushion 29
Mosses 29
Moths
  Eastern Tent Caterpillar 77
  Nessus Sphinx 77
  Mourning Cloak 76
Mouse, Deer 112
Mouse and Rat Family 112-113
Mullein, Common 52
Musclewood 34
Mushrooms 24
Muskrat, Common 115
Mussel, Zebra 86
Mustard, Garlic 46
Mustard Family 46, 59
Mustela
  erminea 114
  frenata 114
  vison 115
Mycorrhizal 24, 25, 49
Myotis lucifugus 112
Nerodia sipedon 95
Nesovitrea binneyana 85
Nettle 63
Nettle Family 63
New World Porcupine Family 116
Newt, Eastern 88
Newt Family 88
Nightshade, Bittersweet 59
Notophthalmus viridescens 88
Nursery Spider Family 79
Nuthatch, White-Breasted 103
Nuthatch Family 103
Nymphalis antiopa 76
Oak
  Black 33
  Northern Red 33
  White 32
Odocoileus virginianus 118
Old World Sparrow Family 108
Oleaster Family 40
Olive Family 39
Ondatra zibethicus 115
Onoclea sensibilis 28
Opiliones Order 80
Opossum, Virginia 116
Orb Weaver, Marbled 78
Orb Weaver Family 78
Orchid Family 51
Oriole, Baltimore 109
Ostrya virginiana 34
Otter, River 117
Otus asio 101
Ovenbird 107

Owl
  Barred 101
  Eastern Screech 101
  Great Horned 101
Owl Family 101
Oxalis sp. 51
Oxidus spp. 87
Oxychilus alliarius 84
Paleontological Research Institution 121
Papilio
  glaucus 95
  polyxenes 94
Parasitic 24, 25, 49
Parsley Family 48, 52
Parsnip, Wild 52
Parthenocissus quinquefolia 63
Partridge Family 110
Passer domesticus 108
Pastinaca sativa 52
Pawpaw 39
Pea Family 40, 48, 52, 56, 59
Pear 35
Peeper, Spring 91
Periwinkle, Common 58
Peromyscus maniculatus 112
Philaenus spumarius 68
Phleum pratense 63
Photuris sp. 70
Phytolacca americana 50
Picea
  abies 30
  glauca 31
  rubens 31
Picoides
  pubescens 102
  villosus 102
Pieris rapae 75
Pigeon and Dove Family 110
Pillar, Glossy 83
Pillar Snail Family 83
Pine
  Eastern White 32
  Red 31
  Scotch 31
Pine Family 30-32
Pink, Deptford 57
Pink Family 57
Pinus
  resinosa 31
  strobus 32
  sylvestris 31
Piranga olivacea 107
Pisaurina mira 79
Pit Viper Family 96
Plainnose Bat Family 112

Plethodon
   cinereus 90
   glutinosus 89
Plover Family 99
Podophyllum peltatum 45
Poecile atricapillus 104
Pokeweed 50
Pokeweed Family 50
Polygala, Fringed 55
Polygala paucifolia 55
Polygonatum biflorum 46
Polypore Family 24
Polystichum acrostichoides 29
Popillia japonica 70
Poppy Family 44, 45
Populus
   deltoides 36
   grandidentata 36
   tremuloides 37
Porcellia sp. 87
Porcupine 116
Potentilla simplex 50
Preventorium 121
Primrose Family 52
Procyon lotor 116
Prowling Spider Family 80
Prunella vulgaris 60
Prunus serotina 36
Pterophylla camellifolia 66
Puffball, Stump 25
Puffball Family, True 25
Punctum minutissimum 84
Purple, Red-Spotted 76
Purslane Family 54
Pyrus communis 35
Queen Anne's Lace 48
Quercus
   alba 32
   rubra 33
   velutina 33
Quiscalus quiscula 109
Raccoon 116
Rana
   catesbeiana 91
   clamitans 90
   palustris 90
   pipiens 90
   sylvatica 90
Ranunculus acris 51
Raspberry
   Purple-flowering 57
   Red 47
Rattlesnake, Timber 96
Redbud, Eastern 40
Reptiles 94-97
Rhamnus cathartica 40

Rhus hirta (typhina) 41
Robin, American 105
Robinia pseudoacacia 38
Rocktripe, Giant 26
Rosa multiflora 42
Rose, Multiflora 42
Rose Family 35, 36, 42, 46, 47, 50, 52, 57
Rubus
   idaeus 47
   odoratus 57
Rudbeckia hirta 53
St. Johnswort, Common 53
St. Johnswort Family 53
Salamander
   Blue-spotted 89
   Four-toed 90
   Jefferson 89
   Mountain Dusky 90
   Northern Dusky 90
   Northern Red-backed 90
   Northern Slimy 89
   Northern Two-lined 89
   Spotted 89
   Spring 90
Salix babylonica 36
Sambucus canadensis 42
Sandpiper Family 99
Sanguinaria canadensis 44
Saprophytic 24
Scarab Beetle Family 70
Sciurus carolinensis 114
Scolopax minor 99
Scolopocryptops spp. 87
Scutigera spp. 87
Seiurus aurocapillus 107
Setaria glauca 63
Shrew, Northern Short-tailed 113
Shrew Family 113
Sialia 104
Sisyrinchium sp. 59
Sitta carolinensis 103
Skink Family 94
Skipper, Silver-spotted 74
Skipper Family 74
Skunk, Striped 115
Slug, Gray Field 82
Slug Family 82
Smilacina racemosa 46
Smith Woods 48
Snails 22, 82-84
Snakeroot, White 50
Snake
   Black Rat 95
   Brown 96
   Common Garter 96

Eastern Milk   95
Northern Ringneck   94
Northern Water   95
Red-belly   96
Ribbon   96
Smooth Green   95
Snapdragon Family   52
Snapping Turtle Family   97
Soils   17
Solanum dulcamara   59
Solidago spp.   53
Solomon's Seal   46
False   46
Sparrow
   House   108
   Song   107
Speyeria cybele   76
Sphinx Moth Family   77
Spider
   Goldenrod Crab   79
   Grass   79
   Nursery Web   79
   Wolf   79
   Yellow Sac   80
Spirobolus spp.   87
Spittlebug, Meadow   68
Spot, Small   84
Spot Snail Family   84
Spring Beauty   54
Spruce
   Norway   30
   Red   31
   White   31
Squamata   94
Squirrel
   Eastern Gray   114
   Flying   114
   Red   114
Squirrel Family   113, 114, 116
Staphylea trifolia   42
Starling, European   106
Starling Family   106
Stinkbug Family   69
Storeria dekayi   96
Strawberry
   Wild   46
   Wood   47
Strix varia   101
Sturnus vulgaris   106
Succinea putris   84
Sulfur Shelf   24
Sulpher, Orange   75
Sumac, Staghorn   41
Supercoil Family   84
Susan, Black-eyed   53

Swallow
   Barn   103
   Tree   102
Swallow Family   102, 103
Swallowtail
   Black   74
   Eastern Tiger   75
Swallowtail Family   74, 75
Swift, Chimney   110
Swift Family   110
Sylvilagus floridanus   115
Sympetrum vicinum   65
Symphyotrichum novae-angliae   60
Symplocarpus foetidus   61
Tachycineta bicolor   102
Tamias striatus   113
Tamiasciurus hudsonicus   114
Tanager, Scarlet   107
Tanager Family   107
Taraxacum officinale   51
Taughannock State Park   18, 22
Taxostoma rufum   105
Teasel   58
Teasel Family   58
Tenodera aridifolia   66
Tent Caterpillar Moth Family   77
Testudines   94
Thalictrum dioicum   61
Thamnophis
   sauritus   96
   sirtalis   96
Thelypteris noveboracensis   28
Thistle, Bull   57
Thoroughwort, Hyssop-leaved   50
Thrasher, Brown   106
Thrush, Wood   105
Thrush Family   104-105
Thuja occidentalis   30
Tibicen sp.   67
Tick, Deer   80
Tiger Beetle Family   69
Tiger Moth Family   77
Tilia americana   34
Timothy   63
Titmouse, Tufted   104
Toad, American   91
Toad Family, True   91
Toads   91
Tomato Family   59
Toothwort   46
Topography   17
Touch-me-not Family   53
Toxicodendron radicans   61
Trametes versicolor   24
Treefrog, Gray   91
Treefrog Family   91

Trees   30-42
Trefoil, Bird's-foot   52
Tremella mesenterica   25
Trifolium pretense   56
Trillium
   Large-flowered (White)   45
   Red   54
Trillium
   erectum   54
   grandiflorum   45
Trilobites   20, 64
Troglodytes aedon   104
Trout-Lily   50
Tsuga canadensi   32
Tulip Tree   35
Turdus migratorius   105
Turkey, Wild   110
Turkey Tail   24
Turtle
   Common Snapping   97
   Painted   97
   Spotted   97
   Wood   97
Twinleaf   45
Typha latifolia   62
Ulmus rubra   34
Umbilicaria mammulata   26
Urnula craterium   26
Urocyon cinereoargenteus   117
Ursus americanus   118
Urtica sp.   63
Vaccinium corymbosum   39
Veery   105
Verbascum thapsus   52
Verbena hastata   52
Vervain, Blue   60
Vervain Family   60
Vespula
   maculata   72
   maculifrons   72
Vetch
   Cow   59
   Crown   56
Viceroy   77
Vicia cracca   59
Vinca minor   58
Viola sororia   58
Violet, Common Blue   58
Violet Family   58
Vitus sp.   62
Vole, Meadow   113
Vulpes vulpes   117
Vulture, Turkey   100
Walkingstick, Northern   66
Walkingstick Family   66
Walnut, Black   38

Walnut Family   38
Warbler, Yellow   106
Wasp Family   72
Water Boatmen   68
Water Strider   64
Waterfowl Family   98
Waterleaf, Virginia   47
Waterleaf Family   47
Wax Cap Family   25
Weasel
   Long-tailed   114
   Short-tailed   114
Weasel Family   114, 115, 117
Webworm, Fall   77
Whirligig Beetle Family   70
White, Cabbage   75
White and Sulphur Family   75
Wildflowers   44-63
Willow, Weeping   36
Willow Family   36-37
Wintergreen Family   49
Witch Hazel   40
Witch's Butter   25
Witch's Hat   25
Wood Fern Family   28, 29
Wood Warbler Family   106, 107
Wood-sorrel, Yellow   51
Wood-sorrel Family   51
Woodchuck   116
Woodcock, American   99
Woodlice   87
Woodlouse Hunter   80
Woodpecker
   Downy   102
   Hairy   102
   Pileated   102
   Red-Bellied   101
Woodpecker Family   101, 102
Worms   86
Wren, House   104
Wren Family   104
Yarrow   49
Yellow Jacket, Eastern   72
Zenaida macroura   110

About the Author

James Dake is an educator, naturalist, and photographer originally from Flushing, Michigan. He served as Nature Center Collaborations Coordinator between the Paleontological Research Institution and the Cayuga Nature Center from 2008 to 2010. He is currently Education Director at Grass River Natural Area in Bellaire, Michigan. The Field Guide to the Cayuga Lake Region was his first published book. He currently lives in Elk Rapids, Michigan.